Astrophysics
for People in a Hurry

Astrophysics
for People in a Hurry

NEIL deGRASSE TYSON

W. W. NORTON & COMPANY

Independent Publishers Since 1923

New York | London

Chapters adapted from "Universe" essays in Natural History *magazine—Chapter 1: March 1998 and September 2003; Chapter 2: November 2000; Chapter 3: October 2003; Chapter 4: June 1999; Chapter 5: June 2006; Chapter 6: October 2002; Chapter 7: July/August 2002; Chapter 8: March 1997; Chapter 9: December 2003/January 2004; Chapter 10: October 2001; Chapter 11: February 2006; Chapter 12: April 2007.*

For information about permission to reproduce selections from this book, write to Permissions, W. W. Norton & Company, Inc., 500 Fifth Avenue, New York, NY 10110

For information about special discounts for bulk purchases, please contact W. W. Norton Special Sales at specialsales@wwnorton.com or 800-233-4830

Manufacturing by Quad Graphics, Fairfield
Book design by Chris Welch
Production manager: Anna Oler

ISBN 978-0-393-60939-4

W. W. Norton & Company, Inc.,
500 Fifth Avenue, New York, NY 10110
www.wwnorton.com

W. W. Norton & Company Ltd.,
15 Carlisle Street, London W1D 3BS

2   3   4   5   6   7   8   9   0

For all those who are too busy to read fat books

Yet nonetheless seek a conduit to the cosmos

*

CONTENTS

PREFACE

In recent years, no more than a week goes by without news of a cosmic discovery worthy of banner headlines. While media gatekeepers may have developed an interest in the universe, this rise in coverage likely comes from a genuine increase in the public's appetite for science. Evidence for this abounds, from hit television shows inspired or informed by science, to the success of science fiction films starring marquee actors, and brought to the screen by celebrated producers and directors. And lately, theatrical release biopics featuring important scientists have become a genre unto itself. There's also widespread interest around the world in science festivals, science fiction conventions, and science documentaries for television.

The highest grossing film of all time is by a famous director who set his story on a planet orbiting a distant star. And it features a famous actress who plays an astrobiologist. While most branches of science have ascended in this era, the field of astrophysics persistently rises to the top. I think I know why. At one time or another every one of us has looked up at the night sky and wondered: What does it all mean? How does it all work? And, what is my place in the universe?

If you're too busy to absorb the cosmos via classes, textbooks, or documentaries, and you nonetheless seek a brief but meaningful introduction to the field, I offer you *Astrophysics for People in a Hurry*. In this slim volume, you will earn a foundational fluency in all the major ideas and discoveries that drive our modern understanding of the universe. If I've succeeded, you'll be culturally conversant in my field of expertise, and you just may be hungry for more.

The universe is under no obligation
to make sense to you.

—NDT

Astrophysics
for People in a Hurry

1.

The Greatest Story
Ever Told

The world has persisted many a long
year, having once been set going in
the appropriate motions. From these
everything else follows.

LUCRETIUS, C. 50 BC

In the beginning, nearly fourteen billion years ago, all the space and all the
matter and all the energy of the known
universe was contained in a volume less than
one-trillionth the size of the period that ends
this sentence.

Conditions were so hot, the basic forces of
nature that collectively describe the universe
were unified. Though still unknown how it
came into existence, this sub-pinpoint-size

cosmos could only expand. Rapidly. In what today we call the big bang.

Einstein's general theory of relativity, put forth in 1916, gives us our modern understanding of gravity, in which the presence of matter and energy curves the fabric of space and time surrounding it. In the 1920s, quantum mechanics would be discovered, providing our modern account of all that is small: molecules, atoms, and subatomic particles. But these two understandings of nature are formally incompatible with one another, which set physicists off on a race to blend the theory of the small with the theory of the large into a single coherent theory of quantum gravity. Although we haven't yet reached the finish line, we know exactly where the high hurdles are. One of them is during the "Planck era" of the early universe. That's the interval of time from $t = 0$ up to $t = 10^{-43}$ seconds (one ten-million-trillion-trillion-trillionths of a second) after the beginning, and before the universe grew to 10^{-35} meters (one hundred billion trillion-trillionths of a meter) across. The German physicist Max Planck, after whom these unimaginably small quantities are named, introduced the idea of quantized energy in 1900 and is generally credited as the father of quantum mechanics.

The clash between gravity and quantum mechanics poses no practical problem for the contemporary universe. Astrophysicists apply the tenets and tools of general relativity and quantum mechanics to very different classes of problems. But in the beginning, during the Planck era, the large was small, and we suspect there must have been a kind of shotgun wedding between the two. Alas, the vows exchanged during that ceremony continue to elude us, and so no (known) laws of physics describe with any confidence the behavior of the universe over that time.

We nonetheless expect that by the end of the Planck era, gravity wriggled loose from the other, still unified forces of nature, achieving an independent identity nicely described by our current theories. As the universe aged through 10^{-35} seconds it continued to expand, diluting all concentrations of energy, and what remained of the unified forces split into the "electroweak" and the "strong nuclear" forces. Later still, the electroweak force split into the electromagnetic and the "weak nuclear" forces, laying bare the four distinct forces we have come to know and love: with the weak force controlling radioactive decay, the strong force binding the atomic nucleus, the electromagnetic force binding molecules, and gravity binding bulk matter.

*

A trillionth of a second has passed since the beginning.

*

All the while, the interplay of matter in the form of subatomic particles, and energy in the form of photons (massless vessels of light energy that are as much waves as they are particles) was incessant. The universe was hot enough for these photons to spontaneously convert their energy into matter-antimatter particle pairs, which immediately thereafter annihilate, returning their energy back to photons. Yes, antimatter is real. And we discovered it, not science fiction writers. These transmogrifications are entirely prescribed by Einstein's most famous equation: $E = mc^2$, which is a two-way recipe for how much matter your energy is worth, and how much energy your matter is worth. The c^2 is the speed of light squared—a huge number which, when multiplied by the mass, reminds us how much energy you actually get in this exercise.

Shortly before, during, and after the strong and electroweak forces parted company, the universe was a seething soup of quarks, leptons, and their antimatter siblings, along with bosons, the particles that enable their interactions. None of these particle families is thought to be divisible into anything smaller or more basic, though each comes in several varieties. The ordinary photon is a member of the boson family. The leptons most familiar to the non-physicist are the electron and perhaps the neutrino; and the most familiar quarks are . . . well, there are no familiar quarks. Each of their six subspecies has been assigned an abstract name that serves no real philological, philosophical, or pedagogical purpose, except to distinguish it from the others: *up* and *down*, *strange* and *charmed*, and *top* and *bottom*.

Bosons, by the way, are named for the Indian

scientist Satyendra Nath Bose. The word "lepton" derives from the Greek *leptos*, meaning "light" or "small." "Quark," however, has a literary and far more imaginative origin. The physicist Murray Gell-Mann, who in 1964 proposed the existence of quarks as the internal constituents of neutrons and protons, and who at the time thought the quark family had only three members, drew the name from a characteristically elusive line in James Joyce's *Finnegans Wake*: "Three quarks for Muster Mark!" One thing quarks do have going for them: all their names are simple—something chemists, biologists, and especially geologists seem incapable of achieving when naming their own stuff.

Quarks are quirky beasts. Unlike protons, each with an electric charge of +1, and electrons, with a charge of −1, quarks have fractional charges that come in thirds. And you'll never catch a quark all by itself; it will always be clutching other quarks nearby. In fact, the force that keeps two (or more) of them together actually grows stronger the more you separate them—as if they were attached by some sort of subnuclear rubber band. Separate the quarks enough, the rubber band snaps and the stored energy summons $E = mc^2$ to create a new quark at each end, leaving you back where you started.

During the quark–lepton era the universe was dense enough for the average separation between unattached quarks to rival the separation between attached quarks. Under those conditions, allegiance between adjacent quarks could not be unambiguously established, and they moved freely among themselves, in spite of being collectively bound to one another. The discovery of this state of matter, a kind of quark cauldron, was reported for the first time in 2002 by a team of physicists at the Brookhaven National Laboratories, Long Island, New York.

Strong theoretical evidence suggests that an episode in the very early universe, perhaps during one of the force splits, endowed the universe with a remarkable asymmetry, in which particles of matter barely outnumbered particles of antimatter: by about a billion-and-one to a billion. That small difference in population would hardly get noticed by anybody amid the continuous creation, annihilation, and re-creation of quarks and antiquarks, electrons and anti-electrons (better known as positrons), and neutrinos and antineutrinos. The odd man out had oodles of opportunities to find somebody to annihilate with, and so did everybody else.

But not for much longer. As the cosmos continued to expand and cool, growing larger than the size of our solar system, the temperature dropped rapidly below a trillion degrees Kelvin.

*

A millionth of a second has passed since the beginning.

*

This tepid universe was no longer hot enough or dense enough to cook quarks, and so they all grabbed dance partners, creating a permanent new family of heavy particles called hadrons (from the Greek *hadros*, meaning "thick"). That quark-to-hadron transition soon resulted in the emergence of protons and neutrons as well as other, less familiar heavy particles, all composed of various combinations of quark species. In Switzerland (back on Earth) the European particle physics collaboration[†] uses a large accelerator to collide beams of hadrons in an attempt to re-create these very conditions. This largest machine in the world is sensibly called the Large Hadron Collider.

The slight matter–antimatter asymmetry afflicting the quark–lepton soup now passed to the hadrons, but with extraordinary consequences.

As the universe continued to cool, the amount of energy available for the spontaneous creation of basic particles dropped. During the hadron era, ambient photons could no longer invoke $E = mc^2$ to manufacture quark–antiquark pairs. Not only that, the photons that emerged from all the remaining annihilations lost energy to the ever-expanding universe, dropping below the threshold required to create hadron–antihadron pairs. For every billion annihilations——leaving a billion photons in

† The European Center for Nuclear Research, better known by its acronym, CERN.

their wake—a single hadron survived. Those loners would get to have all the fun: serving as the ultimate source of matter to create galaxies, stars, planets, and petunias.

Without the billion-and-one to a billion imbalance between matter and antimatter, all mass in the universe would have self-annihilated, leaving a cosmos made of photons *and nothing else*—the ultimate let-there-be-light scenario.

✳

By now, one second of time has passed.

✳

The universe has grown to a few light-years across,† about the distance from the Sun to its closest neighboring stars. At a billion degrees, it's still plenty hot—and still able to cook electrons, which, along with their positron counterparts, continue to pop in and out of existence. But in the ever-expanding, ever-cooling universe, their days (seconds, really)

---

† A light-year is the distance light travels in one Earth year—nearly six trillion miles or ten trillion kilometers.

are numbered. What was true for quarks, and true for hadrons, had become true for electrons: eventually only one electron in a billion survives. The rest annihilate with positrons, their antimatter sidekicks, in a sea of photons.

Right about now, one electron for every proton has been "frozen" into existence. As the cosmos continues to cool—dropping below a hundred million degrees—protons fuse with other protons as well as with neutrons, forming atomic nuclei and hatching a universe in which ninety percent of these nuclei are hydrogen and ten percent are helium, along with trace amounts of deuterium ("heavy" hydrogen), tritium (even heavier hydrogen), and lithium.

＊

Two minutes have now passed since the beginning.

＊

For another 380,000 years not much will happen to our particle soup. Throughout these millennia the temperature remains hot enough for electrons to roam free among the photons, batting them to and fro as they interact with one another.

But this freedom comes to an abrupt end when the temperature of the universe falls below 3,000 degrees Kelvin (about half the temperature of the Sun's surface), and all the free electrons combine with nuclei. The marriage leaves behind a ubiquitous bath of visible light, forever imprinting the sky with a record of where all the matter was in that moment, and completing the formation of particles and atoms in the primordial universe.

*

For the first billion years, the universe continued to expand and cool as matter gravitated into the massive concentrations we call galaxies. Nearly a hundred billion of them formed, each containing hundreds of billions of stars that undergo thermonuclear fusion in their cores. Those stars with more than about ten times the mass of the Sun achieve sufficient pressure and temperature in their cores to manufacture dozens of elements heavier than hydrogen, including those that compose planets and whatever life may thrive upon them.

These elements would be stunningly useless were they to remain where they formed. But high-mass stars fortuitously explode, scattering their chemically enriched guts throughout the galaxy. After nine billion years of such enrichment, in an undistinguished part of the universe (the outskirts of the Virgo Supercluster) in an undistinguished galaxy (the Milky Way) in an undistinguished region (the Orion Arm), an undistinguished star (the Sun) was born.

The gas cloud from which the Sun formed contained a sufficient supply of heavy elements to coalesce and spawn a complex inventory of orbiting objects that includes several rocky and gaseous planets, hundreds of thousands of asteroids, and billions of comets. For the first several hundred million years, large quantities of leftover debris in wayward orbits would accrete onto larger bodies. This occurred in the form of high-speed, high-energy impacts, which rendered molten the surfaces of the rocky planets, preventing the formation of complex molecules.

As less and less accretable matter remained in the solar system, planet surfaces began to cool. The one we call Earth formed in a kind of Goldilocks zone around the Sun, where oceans remain largely in liquid form. Had Earth been much closer to the Sun, the oceans would have evaporated. Had Earth been much farther away, the oceans would have frozen. In either case, life as we know it would not have evolved.

Within the chemically rich liquid oceans, by a mechanism yet to be discovered, organic molecules transitioned to self-replicating life. Dominant in this primordial soup were simple anaerobic bacteria—life that thrives in oxygen-empty environments but excretes chemically potent oxygen as one of its by-products. These early, single-celled organisms unwittingly transformed Earth's carbon dioxide-rich atmosphere into one with sufficient oxygen to allow aerobic organisms to emerge and dominate the oceans and land. These same oxygen atoms, normally found in pairs (O_2), also combined in threes to form ozone (O_3) in the upper atmosphere, which serves as a shield that protects Earth's surface from most of the Sun's molecule-hostile ultraviolet photons.

We owe the remarkable diversity of life on Earth, and we presume elsewhere in the universe, to the cosmic abundance of carbon and the countless number of simple and complex molecules that contain it. There's no doubt about it: more varieties of carbon-based molecules exist than all other kinds of molecules combined.

But life is fragile. Earth's occasional encounters with large, wayward comets and asteroids, a formerly common event, wreaks intermittent havoc upon our ecosystem. A mere sixty-five million years ago (less than two percent of Earth's past), a ten-trillion-ton asteroid hit what is now the Yucatan Peninsula and obliterated more than seventy percent of Earth's flora and fauna—including all the famous outsized dinosaurs. Extinction. This ecological catastrophe enabled our mammal ancestors to fill freshly vacant niches, rather than continue to serve as hors d'oeuvres for *T. rex*. One big-brained branch of these mammals, that which we call primates, evolved a genus and species (*Homo sapiens*) with sufficient intelligence to invent methods and tools of science—and to deduce the origin and evolution of the universe.

*

What happened before all this? What happened before the beginning?

Astrophysicists have no idea. Or, rather, our most creative ideas have little or no grounding in experimental science. In response, some religious people assert, with a tinge of righteousness, that *something* must have started it all: a force greater than all others, a source from which everything issues. A prime mover. In the mind of such a person, that something is, of course, God.

But what if the universe was always there, in a state or condition we have yet to identify—a multiverse, for instance, that continually births universes? Or what if the universe just popped into existence from nothing? Or what if everything we know and love were just a computer simulation rendered for entertainment by a superintelligent alien species?

These philosophically fun ideas usually satisfy nobody. Nonetheless, they remind us that ignorance is the natural state of mind for a research scientist. People who believe they are ignorant of nothing have neither looked for, nor stumbled upon, the boundary between what is known and unknown in the universe.

What we do know, and what we can assert without further hesitation, is that the universe had a beginning. The universe continues to evolve. And yes, every one of our body's atoms is traceable to the big bang and to the thermo-nuclear furnaces within high-mass stars that exploded more than five billion years ago.

We are stardust brought to life, then empow-ered by the universe to figure itself out—and we have only just begun.

2.

On Earth as in the Heavens

Until Sir Isaac Newton wrote down the universal law of gravitation, nobody had any reason to presume that the laws of physics at home were the same as everywhere else in the universe. Earth had earthly things going on and the heavens had heavenly things going on. According to Christian teachings of the day, God controlled the heavens, rendering them unknowable to our feeble mortal minds. When Newton breached this philosophical barrier by rendering all motion comprehensible and predictable, some theologians criticized him for leaving nothing for the Creator to do. Newton had figured out that the force of gravity pulling ripe apples from their orchards also guides tossed objects along their curved trajectories and directs the Moon in its orbit around Earth. Newton's law of gravity also guides planets, asteroids, and comets in their orbits around the Sun and keeps hundreds of billions of stars in orbit within our Milky Way galaxy.

This universality of physical laws drives scientific discovery like nothing else. And gravity was just the beginning. Imagine the excitement among nineteenth-century astronomers when laboratory prisms, which break light beams into a spectrum of colors, were first turned to the Sun. Spectra are not only beautiful, but contain oodles of information about the light-emitting object, including its temperature and composition. Chemical elements reveal themselves by their unique patterns of light or dark bands that cut across the spectrum. To people's delight and amazement, the chemical signatures on the Sun were identical to those in the laboratory. No longer the exclusive tool of chemists, the prism showed that as different as the Sun is from Earth in size, mass, temperature, location, and appearance,

we both contain the same stuff: hydrogen, carbon, oxygen, nitrogen, calcium, iron, and so
forth. But more important than our laundry
list of shared ingredients was the recognition that the laws of physics prescribing the
formation of these spectral signatures on the
Sun were the same laws operating on Earth,
ninety-three million miles away.

So fertile was this concept of universality
that it was successfully applied in reverse. Further analysis of the Sun's spectrum revealed
the signature of an element that had no known
counterpart on Earth. Being of the Sun, the
new substance was given a name derived
from the Greek word *helios* ("the Sun"), and
was only later discovered in the lab. Thus,
helium became the first and only element in
the chemist's Periodic Table to be discovered
someplace other than Earth.

Okay, the laws of physics work in the solar
system, but do they work across the galaxy?
Across the universe? Across time itself? Step
by step, the laws were tested. Nearby stars also
revealed familiar chemicals. Distant binary
stars, bound in mutual orbit, seem to know all

about Newton's laws of gravity. For the same reason, so do binary galaxies.

And, like the geologist's stratified sediments, which serve as a timeline of earthly events, the farther away we look in space, the further back in time we see. Spectra from the most distant objects in the universe show the same chemical signatures that we see nearby in space and in time. True, heavy elements were less abundant back then—they are manufactured primarily in subsequent generations of exploding stars—but the laws describing the atomic and molecular processes that created these spectral signatures remain intact. In particular, a quantity known as the fine-structure constant, which controls the basic fingerprinting for every element, must have remained unchanged for billions of years.

Of course, not all things and phenomena in the cosmos have counterparts on Earth. You've probably never walked through a cloud of glowing million-degree plasma, and I'd bet you've never greeted a black hole on the street. What matters is the universality of the physical laws that describe them. When spectral analysis was first applied to the light emitted by interstellar nebulae, a signature was discovered that, once again, had no counterpart on Earth. At the time, the Periodic Table of Elements had no obvious place for a new element to fit. In response, astrophysicists invented the name "nebulium" as a place-holder, until they could figure out what was going on. Turned out that in space, gaseous nebulae are so rarefied that atoms go long stretches without colliding. Under these conditions, electrons can do things within atoms that had never before been seen in Earth labs. Nebulium was simply the signature of ordinary oxygen doing extraordinary things.

This universality of physical laws tells us that if we land on another planet with a thriving alien civilization, they will be running on the same laws that we have discovered and tested here on Earth—even if the aliens harbor different social and political beliefs. Furthermore, if you wanted to talk to the aliens, you can bet they don't speak English or French or even Mandarin. Nor would you know whether shaking their hands—

if indeed their outstretched appendage is a hand—would be considered an act of war or of peace. Your best hope is to find a way to communicate using the language of science.

Such an attempt was made in the 1970s with *Pioneer 10* and *11* and *Voyager 1* and *2*. All four spacecraft were endowed with enough energy, after gravity assists from the giant planets, to escape the solar system entirely.

Pioneer wore a golden etched plaque that showed, in scientific pictograms, the layout of our solar system, our location in the Milky Way galaxy, and the structure of the hydrogen atom. *Voyager* went further and also included a gold record album containing diverse sounds from mother Earth, including the human heartbeat, whale "songs," and musical selections from around the world, including the works of Beethoven and Chuck Berry. While this humanized the message, it's not clear whether alien ears would have a clue what they were listening to—assuming they have ears in the first place. My favorite parody of this gesture was a skit on NBC's *Saturday Night Live*, shortly after the *Voyager* launch, in which they showed a written reply from the aliens who recovered the spacecraft. The note simply requested, "Send more Chuck Berry."

Science thrives not only on the universality of physical laws but also on the existence and persistence of physical constants. The constant of gravitation, known by most scientists as "big G," supplies Newton's equation of gravity with the measure of how strong the force will be. This quantity has been implicitly tested for variation over eons. If you do the math, you can determine that a star's luminosity is steeply dependent on big G. In other words, if big G had been even slightly different in the past, then the energy output of the Sun would have been far more variable than anything the biological, climatological, or geological records indicate.

Such is the uniformity of our universe.

*

Among all constants, the speed of light is the most famous. No matter how fast you go, you will never overtake a beam of light. Why not? No experiment ever conducted has ever revealed an object of any form reaching the speed of light. Well-tested laws of physics predict and account for that fact. I know these statements sound closed-minded. Some of the most bone-headed, science-based proclamations in the past have underestimated the ingenuity of inventors and engineers: "We will never fly." "Flying will never be commercially feasible." "We will never split the atom." "We will never break the sound barrier." "We will never go to the Moon." What they have in common is that no established law of physics stood in the their way.

The claim "We will never outrun a beam of light" is a qualitatively different prediction. It flows from basic, time-tested physical principles. Highway signs for interstellar travelers of the future will justifiably read:

> The Speed of Light:
> It's Not Just a Good Idea
> It's the Law.

Unlike getting caught speeding on Earth roads, the good thing about the laws of physics is that they require no law enforcement agencies to maintain them, although I did once own a geeky T-shirt that proclaimed, "OBEY GRAVITY."

All measurements suggest that the known fundamental constants, and the physical laws that reference them, are neither time-dependent nor location-dependent. They're truly constant and universal.

*

Many natural phenomena manifest multiple physical laws operating at once. This fact often complicates the analysis and, in most cases, requires high-performance computing to calculate what's going on and to keep track of important parameters. When comet Shoemaker-Levy 9 plunged into Jupiter's gas-rich atmosphere in July 1994, and then exploded, the most accurate computer model combined the laws of fluid mechanics, thermodynamics, kinematics, and gravitation. Climate and weather represent other leading examples of complicated (and difficult-to-predict) phenomena. But the basic laws governing them are still at work. Jupiter's Great Red Spot, a raging anticyclone that has been going strong for at least 350 years, is driven by identical physical processes that generate storms on Earth and elsewhere in the solar system.

Another class of universal truths is the conservation laws, where the amount of some measured quantity remains unchanged *no matter what*. The three most important are the conservation of mass and energy, the conservation of linear and angular momentum, and the conservation of electric charge. These laws are in evidence on Earth, and everywhere we have thought to look—from the domain of particle physics to the large-scale structure of the universe.

In spite of this boasting, all is not perfect in paradise. It happens that we cannot see, touch, or taste the source of eighty-five percent of the gravity we measure in the universe. This mysterious *dark matter*, which remains undetected except for its gravitational pull on matter we see, may be composed of exotic particles that we have yet to discover or identify. A small minority of astrophysicists, however, are unconvinced and have suggested that there is no dark matter—you just need to modify Newton's law of gravity. Simply add a few components to the equations and all will be well.

Perhaps one day we will learn that Newton's gravity indeed requires adjustment. That'll be okay. It has happened once before. Einstein's 1916 general theory of relativity expanded on the principles of Newton's gravity in a way that also applied to objects of extremely high mass. Newton's law of gravity breaks down in this expanded realm, which was unknown to him. The lesson here is that our confidence flows through the range of conditions over which a law has been tested and verified. The broader that range, the more potent and powerful the law becomes in describing the cosmos. For ordinary household gravity, Newton's law works just fine. It got us to the Moon and returned us safely to Earth in 1969. For black holes and the large-scale structure of the universe, we need general relativity. And if you insert low mass and low speeds into Einstein's equations they literally (or, rather, mathematically) become Newton's equations—all good reasons to develop confidence in our understanding of all we claim to understand.

*

To the scientist, the universality of physical laws makes the cosmos a marvelously simple place. By comparison, human nature—the psychologist's domain—is infinitely more daunting. In America, local school boards vote on subjects to be taught in the classroom. In some cases, votes are cast according to the whims of cultural, political, or religious tides. Around the world, varying belief systems lead to political differences that are not always resolved peacefully. The power and beauty of physical laws is that they apply everywhere, whether or not you choose to believe in them.

In other words, after the laws of physics, everything else is opinion.

Not that scientists don't argue. We do. A lot. But when we do, we typically express opinions about the interpretation of insufficient or ratty data on the bleeding frontier of our knowledge. Wherever and whenever a physical law can be invoked in the discussion, the debate is guaranteed to be brief: No, your idea for a perpetual motion machine will never work; it violates well-tested laws of thermo-dynamics. No, you can't build a time machine that will enable you to go back and kill your mother before you were born—it violates cau-sality laws. And without violating momentum laws, you cannot spontaneously levitate and hover above the ground, whether or not you are seated in the lotus position.[†]

Knowledge of physical laws can, in some cases, give you the confidence to confront surly people. A few years ago I was having a hot-cocoa nightcap at a dessert shop in Pasadena, California. Ordered it with whipped cream, of course. When it arrived at the table, I saw no trace of the stuff. After I told the waiter that my cocoa had no whipped cream, he asserted I couldn't see it because it sank to the bottom. But whipped cream has low density, and floats on all liquids that humans consume. So I offered the waiter two possible explanations: either somebody forgot to add the whipped cream to my hot cocoa or the

---

[†] You could, in principle, perform this stunt if you managed to let forth a powerful and sustained exhaust of flatulence.

universal laws of physics were different in his restaurant. Unconvinced, he defiantly brought over a dollop of whipped cream to demonstrate his claim. After bobbing once or twice the whipped cream rose to the top, safely afloat.

What better proof do you need of the universality of physical law?

3.

Let There Be Light

After the big bang, the main agenda of the cosmos was expansion, ever diluting the concentration of energy that filled space. With each passing moment the universe got a little bit bigger, a little bit cooler, and a little bit dimmer. Meanwhile, matter and energy co-inhabited a kind of opaque soup, in which free-range electrons continually scattered photons every which way.

For 380,000 years, things carried on that way.

In this early epoch, photons didn't travel far before encountering an electron. Back then, if your mission had been to see across the universe, you couldn't. Any photon you detected had careened off an electron right in front

of your nose, nano- and picoseconds earlier.[†]
Since that's the largest distance that information can travel before reaching your eyes, the
entire universe was simply a glowing opaque
fog in every direction you looked. The Sun and
all other stars behave this way, too.

As the temperature drops, particles move
more and more slowly. And so right about
then, when the temperature of the universe
first dipped below a red-hot 3,000 degrees
Kelvin, electrons slowed down just enough to
be captured by passing protons, thus bringing full-fledged atoms into the world. This
allowed previously harassed photons to be set
free and travel on uninterrupted paths across
the universe.

This "cosmic background" is the incarna-
tion of the leftover light from a dazzling, siz-
zling early universe, and can be assigned a
temperature, based on what part of the spectrum the dominant photons represent. As the
cosmos continued to cool, the photons that had

[†] One nanosecond is a billionth of a second. One picosec-
ond is a trillionth of a second.

been born in the visible part of the spectrum lost energy to the expanding universe and eventually slid down the spectrum, morphing into infrared photons. Although the visible light photons had become weaker and weaker, they never stopped being photons.

What's next on the spectrum? Today, the universe has expanded by a factor of 1,000 from the time photons were set free, and so the cosmic background has, in turn, cooled by a factor of 1,000. All the visible light photons from that epoch have become 1/1,000th as energetic. They're now microwaves, which is where we derive the modern moniker "cosmic microwave background," or CMB for short. Keep this up and fifty billion years from now astrophysicists will be writing about the cosmic radiowave background.

When something glows from being heated, it emits light in all parts of the spectrum, but will always peak somewhere. For household lamps that still use glowing metal filaments, the bulbs all peak in the infrared, which is the single greatest contributor to their inefficiency as a source of visible light. Our senses detect infrared only in the form of warmth on our skin. The LED revolution in advanced lighting technology creates pure visible light without wasting wattage on invisible parts of the spectrum. That's how you can get crazy-sounding sentences on the packaging like: "7 Watts LED replaces 60 Watts Incandescent."

Being the remnant of something that was once brilliantly aglow, the CMB has the profile we expect of a radiant but cooling object: it peaks in one part of the spectrum but radiates in other parts of the spectrum as well. In this case, besides peaking in microwaves, the CMB also gives off some radio waves and a vanishingly small number of photons of higher energy.

In the mid-twentieth century, the subfield of cosmology—not to be confused with cosmetology—didn't have much data. And where data are sparse, competing ideas abound that are clever and wishful. The existence of the CMB was predicted by the Russian-born American physicist George Gamow and colleagues during the 1940s. The foundation of these ideas came from the 1927 work of the

Belgian physicist and priest Georges Lemaître, who is generally recognized as the "father" of big bang cosmology. But it was American physicists Ralph Alpher and Robert Herman who, in 1948, first estimated what the temperature of the cosmic background ought to be. They based their calculations on three pillars: 1) Einstein's 1916 general theory of relativity; 2) Edwin Hubble's 1929 discovery that the universe is expanding; and 3) atomic physics developed in laboratories before and during the Manhattan Project that built the atomic bombs of World War II.

Herman and Alpher calculated and proposed a temperature of 5 degrees Kelvin for the universe. Well, that's just plain wrong. The precisely measured temperature of these microwaves is 2.725 degrees, sometimes written as simply 2.7 degrees, and if you're numerically lazy, nobody will fault you for rounding the temperature of the universe to 3 degrees.

Let's pause for a moment. Herman and Alpher used atomic physics freshly gleaned in a lab, and applied it to hypothesized conditions in the early universe. From this, they extrapolated billions of years forward, calculating what temperature the universe should be today. That their prediction even remotely approximated the right answer is a stunning triumph of human insight. They could have been off by a factor or ten, or a hundred, or they could have predicted something that isn't even there. Commenting on this feat, the American astrophysicist J. Richard Gott noted, "Predicting that the background existed and then getting its temperature correct to within a factor of 2, was like predicting that a flying saucer 50 feet wide would land on the White House lawn, but instead, a flying saucer 27 feet wide actually showed up."

*

The first direct observation of the cosmic microwave background was made inadvertently in 1964 by American physicists Arno Penzias and Robert Wilson of Bell Telephone Laboratories, the research branch of AT&T. In the 1960s everyone knew about microwaves, but almost no one had the tech-

nology to detect them. Bell Labs, a pioneer
in the communications industry, developed
a beefy, horn-shaped antenna for just that
purpose.

But first, if you're going to send or receive a
signal, you don't want too many sources con-
taminating it. Penzias and Wilson sought to
measure background microwave interference
to their receiver, to enable clean, noise-free
communication within this band of the spectrum. They were not cosmologists. They were
techno-wizards honing a microwave receiver,
and unaware of the Gamow, Herman, and
Alpher predictions.

What Penzias and Wilson were decidedly
not looking for was the cosmic microwave
background; they were just trying to open a
new channel of communication for AT&T.

Penzias and Wilson ran their experiment,
and subtracted from their data all the known
terrestrial and cosmic sources of interference
they could identify. But one part of the signal didn't go away, and they just couldn't figure
out how to eliminate it. Finally they looked
inside the dish and saw pigeons nesting there.

They were worried that a white dielectric substance (pigeon poop) might be responsible for
the signal, because they detected it no matter what direction the detector pointed. After
cleaning out the dielectric substance, the
interference dropped a little bit, but a leftover
signal remained. The paper they published in
1965 was all about this unaccountable "excess
antenna temperature."[†]

Meanwhile, a team of physicists at Princeton, led by Robert Dicke, was building a detector specifically to find the CMB. But they
didn't have the resources of Bell Labs, so their
work went a little slower. And the moment
Dicke and his colleagues heard about Penzias
and Wilson's work, the Princeton team knew
exactly what the observed excess antenna temperature was. Everything fit: especially the
temperature itself, and that the signal came
from every direction in the sky.

In 1978, Penzias and Wilson won the Nobel

[†] A. A. Penzias and R. W. Wilson, "A Measurement of
Excess Antenna Temperature at 4080 Mc/s," *Astrophysical
Journal* 142 (1965): 419–21.

Prize for their discovery. And in 2006, American astrophysicists John C. Mather and George F. Smoot would share the Nobel Prize for observing the CMB over a broad range of the spectrum, bringing cosmology from a nursery of clever but untested ideas into the realm of a precision, experimental science.

*

Because light takes time to reach us from distant places in the universe, if we look out in deep space we actually see eons back in time. So if the intelligent inhabitants of a galaxy far, far away were to measure the temperature of the cosmic background radiation at the moment captured by our gaze, they should get a reading higher than 2.7 degrees, because they are living in a younger, smaller, hotter universe than we are.

Turns out you can actually test this hypothesis. The molecule cyanogen CN (once used on convicted murderers as the active component of the gas administered by their executioners) gets excited by exposure to microwaves. If the microwaves are warmer than the ones in our CMB, they excite the molecule a little more. In the big bang model, the cyanogen in distant, younger galaxies gets bathed in a warmer cosmic background than the cyanogen in our own Milky Way galaxy. And that's exactly what we observe.

You can't make this stuff up.

Why should any of this be interesting? The universe was opaque until 380,000 years after the big bang, so you could not have witnessed matter taking shape even if you'd been sitting front-row center. You couldn't have seen where the galaxy clusters and voids were starting to form. Before anybody could have seen anything worth seeing, photons had to travel, unimpeded, across the universe, as carriers of this information.

The spot where each photon began its cross-cosmos journey is where it had smacked into the last electron that would ever stand in its way—the "point of last scatter." As more and more photons escape unsmacked, they create an expanding "surface" of last scatter, some 120,000 years deep. That surface is where all the atoms in the universe were born: an electron joins an atomic nucleus, and a little pulse of energy in the form of a photon soars away into the wild red yonder.

By then, some regions of the universe had already begun to coalesce by the gravitational attraction of their parts. Photons that last scattered off electrons in these regions developed a different, slightly cooler profile than those scattering off the less sociable electrons sitting in the middle of nowhere. Where matter accumulated, the strength of gravity grew, enabling more and more matter to gather. These regions seeded the formation of galaxy superclusters while other regions were left relatively empty.

When you map the cosmic microwave background in detail, you find that it's not completely smooth. It's got spots that are slightly hotter and slightly cooler than average. By studying these temperature variations in the CMB—that is to say, by studying patterns in the surface of last scatter—we can infer what the structure and content of the matter was in the early universe. To figure out how galaxies and clusters and superclusters arose, we use our best probe, the CMB—a potent time capsule that empowers astrophysicists to reconstruct cosmic history in reverse. Studying its patterns is like performing some sort of cosmic phrenology, as we analyze the skull bumps of the infant universe.

When constrained by other observations of the contemporary and distant universe, the CMB enables you to decode all sorts of fundamental cosmic properties. Compare the distribution of sizes and temperatures of the warm and cool areas and you can infer how strong the force of gravity was at the time and how quickly matter accumulated, allowing you to then deduce how much ordinary matter, dark matter, and dark energy there is in the universe. From here, it's then straightforward to tell whether or not the universe will expand forever.

*

Ordinary matter is what we are all made of. It has gravity and interacts with light. Dark matter is a mysterious substance that has gravity but does not interact with light in any known way. Dark energy is a mysterious pressure in the vacuum of space that acts in the opposite direction of gravity, forcing the universe to expand faster than it otherwise would.

What our phrenological exam says is that we understand how the universe behaved, but that most of the universe is made of stuff about which we are clueless. Our profound areas of ignorance notwithstanding, today, as never before, cosmology has an anchor, because the CMB reveals the portal through which we all walked. It's a point where interesting physics happened, and where we learned about the universe before and after its light was set free.

The simple discovery of the cosmic microwave background turned cosmology into something more than mythology. But it was the accurate and detailed map of the cosmic microwave background that turned cosmology into a modern science. Cosmologists have plenty of ego. How could you not when your job is to deduce what brought the universe into existence? Without data, their explanations were just hypotheses. Now, each new observation, each morsel of data, wields a two-edged sword: it enables cosmology to thrive on the kind of foundation that so much of the rest of science enjoys, but it also constrains theories that people thought up when there wasn't enough data to say whether they were right or wrong.

No science achieves maturity without it.

4.

Between the Galaxies

In the grand tally of cosmic constituents, galaxies are what typically get counted. Latest estimates show that the observable universe may contain a hundred billion of them. Bright and beautiful and packed with stars, galaxies decorate the dark voids of space like cities across a country at night. But just how voidy is the void of space? (How empty is the countryside between cities?) Just because galaxies are in your face, and just because they would have us believe that nothing else matters, the universe may nonetheless contain hard-to-detect things between the galaxies. Maybe those things are more interesting, or more important to the evolution of the universe, than the galaxies themselves.

Our own spiral-shaped galaxy, the Milky Way, is named for its spilled-milk appearance to the unaided eye across Earth's nighttime sky. Indeed, the very word "galaxy" derives from the Greek *galaxias*, "milky." Our pair of nearest-neighbor galaxies, 600,000 light-years distant, are both small and irregularly shaped. Ferdinand Magellan's ship's log identified these cosmic objects during his famous round-the-world voyage of 1519. In his honor, we call them the Large and Small Magellanic Clouds, and they are visible primarily from the Southern Hemisphere as a pair of cloudlike splotches on the sky, parked beyond the stars. The nearest galaxy larger than our own is two million light-years away, beyond the stars that trace the constellation Andromeda. This spiral galaxy, historically dubbed the Great Nebula in Andromeda, is a somewhat more massive and luminous twin of the Milky Way. Notice that the name for each system lacks reference to the existence of stars: Milky Way, Magellanic Clouds, Andromeda Nebula. All three were named before telescopes were invented, so they could not yet be resolved into their stellar constituencies.

*

As detailed in chapter 9, without the benefit of telescopes operating in multiple bands of light we might still declare the space between the galaxies to be empty. Aided by modern detectors, and modern theories, we have probed our cosmic countryside and revealed all manner of hard-to-detect things: dwarf galaxies, runaway stars, runaway stars that explode, million-degree X-ray-emitting gas, dark matter, faint blue galaxies, ubiquitous gas clouds, super-duper high-energy charged particles, and the mysterious quantum vacuum energy. With a list like that, one could argue that all the fun in the universe happens between the galaxies rather than within them.

In any reliably surveyed volume of space, dwarf galaxies outnumber large galaxies by more than ten to one. The first essay I ever wrote on the universe, in the early 1980s, was titled "The Galaxy and the Seven Dwarfs," referring to the Milky Way's diminutive nearby family. Since then, the tally of local dwarf galaxies has been counted in the dozens. While full-blooded galaxies contain hundreds of billions of stars, dwarf galaxies can have as few as a million, which renders them a hundred thousand times harder to detect. No wonder they are still being discovered in front of our noses.

Images of dwarf galaxies that no longer manufacture stars tend to look like tiny, boring smudges. Those dwarfs that do form stars are all irregularly shaped and, quite frankly, are a sorry-looking lot. Dwarf galaxies have three things working against their detection: They are small, and so are easily passed over when seductive spiral galaxies vie for your attention. They are dim, and so are missed in many surveys of galaxies that cut off below a prespecified brightness level. And they have a low density of stars within them, so they offer poor contrast above the glow of surrounding light from Earth's nighttime atmosphere and from other sources. All this is true. But since dwarfs far outnumber "normal" galaxies, perhaps our definition of what is normal needs revision.

You will find most (known) dwarf galaxies hanging out near bigger galaxies, in orbit

around them like satellites. The two Magellanic Clouds are part of the Milky Way's dwarf
family. But the lives of satellite galaxies can
be quite hazardous. Most computer models of
their orbits show a slow decay that ultimately
results in the hapless dwarfs getting ripped
apart, and then eaten, by the main galaxy.
The Milky Way engaged in at least one act
of cannibalism in the last billion years, when
it consumed a dwarf galaxy whose flayed
remains can be seen as a stream of stars orbiting the galactic center, beyond the stars of the
constellation Sagittarius. The system is called
the Sagittarius Dwarf, but should probably
have been named Lunch.

In the high-density environment of clusters,
two or more large galaxies routinely collide and
leave behind a titanic mess: spiral structures
warped beyond all recognition, newly induced
bursts of star-forming regions spawned from
the violent collision of gas clouds, and hundreds
of millions of stars strewn hither and yon having freshly escaped the gravity of both galaxies.
Some stars reassemble to form blobs that could
be called dwarf galaxies. Other stars remain

adrift. About ten percent of all large galaxies show evidence of a major gravitational encounter with another large galaxy—and that rate may be five times higher among galaxies in clusters.

With all this mayhem, how much galactic flotsam permeates intergalactic space, especially within clusters? Nobody knows for sure. The measurement is difficult because isolated stars are too dim to detect individually. We must rely on detecting a faint glow produced by the light of all stars combined. In fact, observations of clusters detect just such a glow between the galaxies, suggesting that there may be as many vagabond, homeless stars as there are stars within the galaxies themselves.

Adding ammo to the discussion, we have found (without looking for them) more than a dozen supernovas that exploded far away from what we presume to be their "host" galaxies. In ordinary galaxies, for every star that explodes in this way, a hundred thousand to a million do not, so isolated supernovas may betray entire populations of undetected stars. Supernovas are stars that have blown themselves to smithereens and, in the process, have temporarily (over several weeks) increased their luminosity a billion-fold, making them visible across the universe. While a dozen homeless supernovas is a relatively small number, many more may await discovery, since most supernova searches systematically monitor known galaxies and not empty space.

*

There's more to clusters than their constituent galaxies and their wayward stars. Measurements made with X-ray-sensitive telescopes reveal a space-filling, intra-cluster gas at tens of millions of degrees. The gas is so hot that it glows strongly in the X-ray part of the spectrum. The very movement of gas-rich galaxies through this medium eventually strips them of their own gas, forcing them to forfeit their capacity to make new stars. That could explain it. But when you calculate the total mass present in this heated gas, for most clusters it exceeds the mass of all galaxies in the cluster by as much as a factor of ten. Worse yet, clusters are overrun by dark matter, which happens to contain up to another factor of ten times the mass of everything else. In other words, if telescopes observed mass rather than light, then our cherished galaxies in clusters would appear as insignificant blips amid a giant spherical blob of gravitational forces.

In the rest of space, outside of clusters, there is a population of galaxies that thrived long ago. As already noted, looking out into the cosmos is analogous to a geologist looking across sedimentary strata, where the history of rock formation is laid out in full view. Cosmic distances are so vast that the travel time for light to reach us can be millions or even billions of years. When the universe was one half its current age, a very blue and very faint species of intermediate-sized galaxy thrived. We see them. They hail from far away. Their blue comes from the glow of freshly formed, short-lived, high-mass, high-temperature, high-luminosity stars. The galaxies are faint not only because they are distant but because the population of luminous stars within them was thin. Like the dinosaurs that came and went, leaving birds as their only modern descendant, the faint blue galaxies no longer exist, but presumably have a counterpart in today's universe. Did all their stars burn out? Have they become invisible corpses strewn throughout the universe? Did they evolve into the familiar dwarf galaxies of today? Or were they all eaten by larger galaxies? We do not know, but their place in the timeline of cosmic history is certain.

With all this stuff between the big galaxies, we might expect some of it to obscure our view of what lies beyond. This could be a problem for the most distant objects in the universe, such as quasars. Quasars are superluminous galaxy cores whose light has typically been traveling for billions of years across space before reaching our telescopes. As extremely distant sources of light, they make ideal guinea pigs for the detection of intervening junk.

Sure enough, when you separate quasar light into its component colors, revealing a spectrum, it's riddled with the absorbing presence of intervening gas clouds. Every known quasar, no matter where on the sky it's found, shows features from dozens of isolated hydrogen clouds scattered across time and space. This unique class of intergalactic object was first identified in the 1980s, and continues to be an active area of astrophysical research. Where did they come from? How much mass do they all contain?

Every known quasar reveals these hydrogen features, so we conclude that the hydrogen clouds are everywhere in the universe. And, as expected, the farther the quasar, the more clouds are present in the spectrum. Some of the hydrogen clouds (less than one percent) are simply the consequence of our line of sight passing through the gas contained in an ordinary spiral or irregular galaxy. You would, of course, expect at least some quasars to fall behind the light of ordinary galaxies that are too distant to detect. But the rest of the absorbers are unmistakable as a class of cosmic object.

Meanwhile, quasar light commonly passes through regions of space that contain monstrous sources of gravity, which wreak havoc on the quasar's image. These are often hard to detect because they may be composed of ordinary matter that is simply too dim and distant, or they may be zones of dark matter, such as what occupies the centers and surrounding regions of galaxy clusters. In either case, where there is mass there is gravity. And where there is gravity there is curved space, according to Einstein's general theory of relativity. And where space is curved it can mimic the curvature of an ordinary glass lens and alter the pathways of light that pass through. Indeed, distant quasars and whole galaxies have been "lensed" by objects that happen to fall along the line of sight to Earth's telescopes. Depending on the mass of the lens itself and the geometry of the line-of-sight alignments, the lensing action can magnify, distort, or even split the background source of light into multiple images, just like fun-house mirrors at arcades.

One of the most distant (known) objects in the universe is not a quasar but an ordinary galaxy, whose feeble light has been magnified significantly by the action of an intervening gravitational lens. We may henceforth need to rely upon these "intergalactic" telescopes to peer where (and when) ordinary telescopes cannot reach, and thus reveal the future holders of the cosmic distance record.

*

Nobody doesn't like intergalactic space, but it can be hazardous to your health if you choose to go there. Let's ignore the fact that you would freeze to death as your warm body tried to reach equilibrium with the 3-degree temperature of the universe. And let's ignore the fact that your blood cells would burst while you suffocated from the lack of atmospheric pressure. These are ordinary dangers. From the department of exotic happenings, intergalactic space is regularly pierced by super-duper high-energy, fast-moving, charged, subatomic particles. We call them cosmic rays. The highest-energy particles among them have a hundred million times the energy that can be generated in the world's largest particle accelerators. Their origin continues to be a mystery, but most of these charged particles are protons, the nuclei of hydrogen atoms, and are moving at 99.9999999999999999999 percent of the speed of light. Remarkably, these single subatomic particles carry enough energy to knock a golf ball from anywhere on a putting green into the cup.

Perhaps the most exotic happenings between (and among) the galaxies in the vacuum of space and time is the seething ocean of virtual particles—undetectable matter and antimatter pairs, popping in and out of existence. This peculiar prediction of quantum physics has been dubbed the "vacuum energy," which manifests as an outward pressure, acting counter to gravity, that thrives in the total absence of matter. The accelerating universe, dark energy incarnate, may be driven by the action of this vacuum energy.

Yes, intergalactic space is, and will forever be, where the action is.

5.

Dark Matter

Gravity, the most familiar of nature's forces, offers us simultaneously the best and the least understood phenomena in nature. It took the mind of the millennium's most brilliant and influential person, Isaac Newton, to realize that gravity's mysterious "action-at-a-distance" arises from the natural effects of every bit of matter, and that the attractive force between any two objects can be described by a simple algebraic equation. It took the mind of the last century's most brilliant and influential person, Albert Einstein, to show that we can more accurately describe gravity's action-at-a-distance as a warp in the fabric of space-time, produced by any combination of matter and energy.

Einstein demonstrated that Newton's theory requires some modification to describe gravity accurately—to predict, for example, how much light rays will bend when they pass by a massive object. Although Einstein's equations are fancier than Newton's, they nicely accommodate the matter that we have come to know and love. Matter that we can see, touch, feel, smell, and occasionally taste.

We don't know who's next in the genius sequence, but we've now been waiting nearly a century for somebody to tell us why the bulk of all the gravitational force that we've measured in the universe—about eighty-five percent of it—arises from substances that do not otherwise interact with "our" matter or energy. Or maybe the excess gravity doesn't come from matter and energy at all, but emanates from some other conceptual thing. In any case, we are essentially clueless. We find ourselves no closer to an answer today than we were when this "missing mass" problem was first fully analyzed in 1937 by the Swiss-American astrophysicist Fritz Zwicky. He taught at the California Institute of Technology for more than forty years, combining his far-ranging insights into the cosmos with a colorful means of expression and an impressive ability to antagonize his colleagues.

Zwicky studied the movement of individual galaxies within a titanic cluster of them, located far beyond the local stars of the Milky Way that trace out the constellation Coma Berenices (the "hair of Berenice," an Egyptian queen in antiquity). The Coma cluster, as we call it, is an isolated and richly populated ensemble of galaxies about 300 million light-years from Earth. Its thousand galaxies orbit the cluster's center, moving in all directions like bees swarming a beehive. Using the motions of a few dozen galaxies as tracers of the gravity field that binds the entire cluster, Zwicky discovered that their average velocity had a shockingly high value. Since larger gravitational forces induce higher velocities in the objects they attract, Zwicky inferred an enormous mass for the Coma cluster. As a reality check on that estimate, you can sum up the masses of each member galaxy that you see. And even though Coma ranks among the largest and most massive clusters in the universe, it does not contain enough visible galaxies to account for the observed speeds Zwicky measured.

How bad is the situation? Have our known laws of gravity failed us? They certainly work within the solar system. Newton showed that you can derive the unique speed that a planet must have to maintain a stable orbit at any distance from the Sun, lest it descend back toward the Sun or ascend to a farther orbit. Turns out, if we could boost Earth's orbital speed to more than the square root of two (1.4142 . . .) times its current value, our planet would achieve "escape velocity," and leave the solar system entirely. We can apply the same reasoning to much larger systems, such as our own Milky Way galaxy, in which stars move in orbits that respond to the gravity from all the other stars; or in clusters of galaxies, where each galaxy likewise feels the gravity from all the other galaxies. In this spirit, amid a page of formulas in his notebook, Einstein wrote a rhyme (more ringingly in German than in this English translation) in honor of Isaac Newton:

Look unto the stars to teach us
How the master's thoughts can reach us
Each one follows Newton's math
Silently along its path.[†]

When we examine the Coma cluster, as Zwicky did during the 1930s, we find that its member galaxies are all moving more rapidly than the escape velocity for the cluster. The cluster should swiftly fly apart, leaving barely a trace of its beehive existence after just a few hundred million years had passed. But the cluster is more than ten billion years old, which is nearly as old as the universe itself. And so was born what remains the longest-standing unsolved mystery in astrophysics.

*

Across the decades that followed Zwicky's work, other galaxy clusters revealed the same problem, so Coma could not be blamed for being peculiar. Then what or who should we

[†] Manuscript note, quoted in Károly Simonyi, *A Cultural History of Physics* (Boca Raton, FL: CRC Press, 2012).

blame? Newton? I wouldn't. Not just yet. His theories had been examined for 250 years and passed all tests. Einstein? No. The formidable gravity of galaxy clusters is still not high enough to require the full hammer of Einstein's general theory of relativity, just two decades old when Zwicky did his research. Perhaps the "missing mass" needed to bind the Coma cluster's galaxies does exist, but in some unknown, invisible form. Today, we've settled on the moniker "dark matter," which makes no assertion that anything is missing, yet nonetheless implies that some new kind of matter must exist, waiting to be discovered.

Just as astrophysicists had come to accept dark matter in galaxy clusters as a mysterious thing, the problem reared its invisible head once again. In 1976, the late Vera Rubin, an astrophysicist at the Carnegie Institution of Washington, discovered a similar mass anomaly within spiral galaxies themselves. Studying the speeds at which stars orbit their galaxy centers, Rubin first found what she expected: within the visible disk of each galaxy, the stars farther from the center move

at greater speeds than stars close in. The farther stars have more matter (stars and gas)
between themselves and the galaxy center,
enabling their higher orbital speeds. Beyond
the galaxy's luminous disk, however, one can
still find some isolated gas clouds and a few
bright stars. Using these objects as tracers of
the gravity field exterior to the most luminous parts of the galaxy, where no more visible matter adds to the total, Rubin discovered
that their orbital speeds, which should now be
falling with increasing distance out there in
Nowheresville, in fact remained high.

These largely empty volumes of space—
the far-rural regions of each galaxy—contain
too little visible matter to explain the anoma-
lously high orbital speeds of the tracers. Rubin
correctly reasoned that some form of dark
matter must lie in these far-out regions, well
beyond the visible edge of each spiral galaxy.
Thanks to Rubin's work, we now call these
mysterious zones "dark matter haloes."

This halo problem exists under our noses,
right in the Milky Way. From galaxy to galaxy and from cluster to cluster, the discrepancy

between the mass tallied from visible objects and the objects' mass estimated from total gravity ranges from a factor of a few up to (in some cases) a factor of many hundreds. Across the universe, the discrepancy averages to a factor of six: cosmic dark matter has about six times the total gravity of all the visible matter.

Further research has revealed that the dark matter cannot consist of ordinary matter that happens to be under-luminous, or nonluminous. This conclusion rests on two lines of reasoning. First, we can eliminate with near-certainty all plausible familiar candidates, like the suspects in a police lineup. Could the dark matter reside in black holes? No, we think that we would have detected this many black holes from their gravitational effects on nearby stars. Could it be dark clouds? No, they would absorb or otherwise interact with light from stars behind them, which bona fide dark matter doesn't do. Could it be interstellar (or intergalactic) rogue planets, asteroids, and comets, all of which produce no light of their own? It's hard to believe that the universe would manufacture six times as much mass in planets as in stars. That would mean six thousand Jupiters for every star in the galaxy, or worse yet, two million Earths. In our own solar system, for example, everything that is not the Sun adds up to less than one fifth of one percent of the Sun's mass.

More direct evidence for the strange nature of dark matter comes from the relative amount of hydrogen and helium in the universe. Together, these numbers provide a cosmic fingerprint left behind by the early universe. To a close approximation, nuclear fusion during the first few minutes after the big bang left behind one helium nucleus for every ten hydrogen nuclei (which are, themselves, simply protons). Calculations show that if most of the dark matter had involved itself in nuclear fusion, there would be much more helium relative to hydrogen in the universe. From this we conclude that most of the dark matter—hence, most of the mass in the universe—does not participate in nuclear fusion, which disqualifies it as "ordinary" matter, whose essence lies in a willingness to participate in the atomic and nuclear forces that shape matter as we know it. Detailed observations of the cosmic microwave background, which allow a separate test of this conclusion, verify the result: Dark matter and nuclear fusion don't mix.

Thus, as best we can figure, the dark matter doesn't simply consist of matter that happens to be dark. Instead, it's something else altogether. Dark matter exerts gravity according to the same rules that ordinary matter follows, but it does little else that might allow us to detect it. Of course, we are hamstrung in this analysis by not knowing what the dark matter is in the first place. If all mass has gravity, does all gravity have mass? We don't know. Maybe there's nothing the matter with the matter, and it's the gravity we don't understand.

*

The discrepancy between dark and ordinary matter varies significantly from one astrophysical environment to another, but it becomes most pronounced for large entities such as galaxies and galaxy clusters. For the smallest objects, such as moons and planets, no discrepancy exists. Earth's surface gravity, for example, can be explained entirely by the stuff that's under our feet. If you are overweight on Earth, don't blame dark matter. Dark matter also has no bearing on the Moon's orbit around Earth, nor on the movements of the planets around the Sun—but as we've already seen, we do need it to explain the motions of stars around the center of the galaxy.

Does a different kind of gravitational physics operate on the galactic scale? Probably not. More likely, dark matter consists of matter whose nature we have yet to divine, and which gathers more diffusely than ordinary matter does. Otherwise, we would detect the gravity of concentrated chunks of dark matter dotting the universe—dark matter comets, dark matter planets, dark matter galaxies. As far as we can tell, that's not the way things are.

What we know is that the matter we have come to love in the universe—the stuff of stars, planets, and life—is only a light frosting on the cosmic cake, modest buoys afloat in a vast cosmic ocean of something that looks like nothing.

*

During the first half million years after
the big bang, a mere eyeblink in the fourteen-
billion-year sweep of cosmic history, matter
in the universe had already begun to coalesce
into the blobs that would become clusters and
superclusters of galaxies. But the cosmos would
double in size during its next half million years,
and continue growing after that. At odds in the
universe were two competing effects: gravity
wants to make stuff coagulate, but the expansion wants to dilute it. If you do the math, you
rapidly deduce that the gravity from ordinary
matter could not win this battle by itself. It
needed the help of dark matter, without which
we would be living—actually not living—in
a universe with no structures: no clusters, no
galaxies, no stars, no planets, no people.

How much gravity from dark matter did it
need? Six times as much as that provided by
ordinary matter itself. Just the amount we measure in the universe. This analysis doesn't tell
us what dark matter is, only that dark matter's
effects are real and that, try as you might, you
cannot credit ordinary matter for it.

*

So dark matter is our frenemy. We have no clue what it is. It's kind of annoying. But we desperately need it in our calculations to arrive at an accurate description of the universe. Scientists are generally uncomfortable whenever we must base our calculations on concepts we don't understand, but we'll do it if we have to. And dark matter is not our first rodeo. In the nineteenth century, for example, scientists measured the energy output of our Sun and showed its effect on our seasons and climate, long before anyone knew that thermonuclear fusion is responsible for that energy. At the time, the best ideas included the retrospectively laughable suggestion that the Sun was a burning lump of coal. Also in the nineteenth century, we observed stars, obtained their spectra, and classified them long before the twentieth-century introduction of quantum physics, which gives us our understanding of how and why these spectra look the way they do.

Unrelenting skeptics might compare the dark matter of today to the hypothetical, now-defunct "aether" proposed in the nineteenth century as the weightless, transparent medium permeating the vacuum of space through which light moved. Until a famous 1887 experiment in Cleveland showed otherwise, performed by Albert Michelson and Edward Morley at Case Western Reserve University, scientists asserted that the aether must exist, even though not a shred of evidence supported this presumption. As a wave, light was thought to require a medium through which to propagate its energy, much as sound requires air or some other substance to transmit its waves. But light turns out to be quite happy traveling through the vacuum of space, devoid of any medium to carry it. Unlike sound waves, which consist of air vibrations, light waves were found to be self-propagating packets of energy requiring no assistance at all.

Dark-matter ignorance differs fundamentally from aether ignorance. The aether was a placeholder for our incomplete understanding, whereas the existence of dark matter derives not from mere presumption but from the observed effects of its gravity on visible matter. We're not inventing dark matter out of thin space; instead, we deduce its existence from observational facts. Dark matter is just as real as the countless exoplanets discovered in orbit around stars other than the Sun, discovered solely through their gravitational influence on their host stars and not from direct measurement of their light.

The worst that can happen is we discover that dark matter does not consist of matter at all, but of something else. Could we be seeing the effects of forces from another dimension? Are we feeling the ordinary gravity of ordinary matter crossing the membrane of a phantom universe adjacent to ours? If so, this could be just one of an infinite assortment of universes that comprise the multiverse. Sounds exotic and unbelievable. But is it any more crazy than the first suggestions that Earth orbits the Sun? That the Sun is one of a hundred billion stars in the Milky Way? Or that the Milky Way is but one of a hundred billion galaxies in the universe?

Even if any of these fantastical accounts prove true, none of it would change the successful invocation of dark matter's gravity in the equations that we use to understand the formation and evolution of the universe.

Other unrelenting skeptics might declare that "seeing is believing"—an approach to life that works well in many endeavors, including mechanical engineering, fishing, and perhaps dating. It's also good, apparently, for residents of Missouri. But it doesn't make for good science. Science is not just about seeing, it's about measuring, preferably with something that's *not* your own eyes, which are inextricably conjoined with the baggage of your brain. That baggage is more often than not a satchel of preconceived ideas, post-conceived notions, and outright bias.

*

Having resisted attempts to detect it directly on Earth for three-quarters of a century, dark matter remains in play. Particle physicists are confident that dark matter consists of a ghostly class of undiscovered particles that interact with matter via gravity, but otherwise interact with matter or light only weakly or not at all.

If you like gambling on physics, this option is a good bet. The world's largest particle accelerators are trying to manufacture dark matter particles amid the detritus of particle collisions. And specially designed laboratories buried deep underground are trying to detect dark matter particles passively, in case they wander in from space. An underground location naturally shields the facility from known cosmic particles that might trip the detectors as dark matter impostors.

Although it all could be much ado about nothing, the idea of an elusive dark matter particle has good precedence. Neutrinos, for instance, were predicted and eventually discovered, even though they interact extremely weakly with ordinary matter. The copious flux of neutrinos from the Sun—two neutrinos for every helium nucleus fused from hydrogen in the Sun's thermonuclear core—exit the Sun unfazed by the Sun itself, travel through the vacuum of space at nearly the speed of light, then pass through Earth as though it does not exist. The tally: night and day, a hundred billion neutrinos from the Sun pass through each thumbnail patch of your body, every second, without a trace of interaction with your body's atoms. In spite of this elusivity, neutrinos are nonetheless stoppable under special circumstances. And if you can stop a particle at all, you've detected it.

Dark matter particles may reveal themselves through similarly rare interactions, or, more amazingly, they might manifest via forces other than the strong nuclear force, weak nuclear force, and electromagnetism. These three, plus gravity, complete the fab four forces of the universe, mediating all interactions between and among all known particles. So the choices are clear. Either dark matter particles must wait for us to discover and to control a new force or class of forces through which their particles interact, or else dark matter particles interact via normal forces, but with staggering weakness.

So, dark matter's effects are real. We just don't know what it is. Dark matter seems not to interact through the strong nuclear force, so it cannot make nuclei. It hasn't been found to interact through the weak nuclear force, something even elusive neutrinos do. It doesn't seem to interact with the electromagnetic force, so it doesn't make molecules and concentrate into dense balls of dark matter. Nor does it absorb or emit or reflect or scatter light. As we've known from the beginning, dark matter does, indeed, exert gravity, to which ordinary matter responds. But that's it. After all these years, we haven't discovered it doing anything else.

For now, we must remain content to carry dark matter along as a strange, invisible friend, invoking it where and when the universe requires it of us.

6.

Dark Energy

As if you didn't have enough to worry about, the universe in recent decades was discovered to wield a mysterious pressure that issues forth from the vacuum of space and that acts opposite cosmic gravity. Not only that, this "negative gravity" will ultimately win the tug-of-war, as it forces the cosmic expansion to accelerate exponentially into the future.

For the most mind-warping ideas of twentieth-century physics, just blame Einstein.

Albert Einstein hardly ever set foot in the laboratory; he didn't test phenomena or use elaborate equipment. He was a theorist who perfected the "thought experiment," in which you engage nature through your imagination, by inventing a situation or model and then working out the consequences of some physical principle. In Germany before World War II, laboratory-based physics far outranked theoretical physics in the minds of most Aryan scientists. Jewish physicists were all relegated to the lowly theorists' sandbox and left to fend for themselves. And what a sandbox that would become.

As was the case for Einstein, if a physicist's model intends to represent the entire universe, then manipulating the model should be tantamount to manipulating the universe itself. Observers and experimentalists can then go out and look for the phenomena predicted by that model. If the model is flawed, or if the theorists make a mistake in their calculations, the observers will uncover a mismatch between the model's predictions and the way things happen in the real universe. That's the first cue for a theorist to return to the proverbial drawing board, by either adjusting the old model or creating a new one.

One of the most powerful and far-reaching theoretical models ever devised, already introduced in these pages, is Einstein's general theory of relativity—but you can call it GR after you get to know it better. Published in 1916, GR outlines the relevant mathematical details of how everything in the universe moves under the influence of gravity. Every few years, lab scientists devise ever more precise experiments to test the theory, only to further extend the envelope of the theory's accuracy. A modern example of this stunning knowledge of nature that Einstein has gifted us, comes from 2016, when gravitational waves were discovered by a specially designed observatory tuned for just this purpose.[†] These waves, predicted by Einstein, are ripples moving at the speed of light across the fabric of space-time, and are generated by severe gravitational disturbances, such as the collision of two black holes.

And that's exactly what was observed. The gravitational waves of the first detection were

[†] The Laser Interferometer Gravitational-Wave Observatory (LIGO), twinned in Hanford, Washington, and Livingston, Louisiana.

generated by a collision of black holes in a galaxy 1.3 billion light-years away, and at a time
when Earth was teeming with simple, single-
celled organisms. While the ripple moved
through space in all directions, Earth would,
after another 800 million years, evolve complex life, including flowers and dinosaurs and
flying creatures, as well as a branch of ver-
tebrates called mammals. Among the mammals, a sub-branch would evolve frontal lobes
and complex thought to accompany them. We
call those primates. A single branch of these
primates would develop a genetic mutation
that allowed speech, and that branch—*Homo
sapiens*—would invent agriculture and civilization and philosophy and art and science.
All in the last ten thousand years. Ultimately,
one of its twentieth-century scientists would
invent relativity out of his head, and predict
the existence of gravitational waves. A century later, technology capable of seeing these
waves would finally catch up with the prediction, just days before that gravity wave,
which had been traveling for 1.3 billion years,
washed over Earth and was detected.

Yes, Einstein was a badass.

*

When first proposed, most scientific models are only half-baked, leaving wiggle room to adjust parameters for a better fit to the known universe. In the Sun-based "heliocentric" universe, conceived by the sixteenth-century mathematician Nicolaus Copernicus, planets orbited in perfect circles. The orbit-the-Sun part was correct, and a major advance on the Earth-based "geocentric" universe, but the perfect-circle part turned out to be a bit off— all planets orbit the Sun in flattened circles called ellipses, and even that shape is just an approximation of a more complex trajectory. Copernicus's basic idea was correct, and that's what mattered most. It simply required some tweaking to make it more accurate.

Yet, in the case of Einstein's relativity, the founding principles of the entire theory require that everything must happen exactly as predicted. Einstein had, in effect, built what looks on the outside like a house of cards, with only two or three simple postulates holding up the entire structure. Indeed, upon learning of a 1931 book entitled *One Hundred Authors Against Einstein,*[†] he responded that if he were wrong, then only one would have been enough.

Therein were sown the seeds of one of the most fascinating blunders in the history of science. Einstein's new equations of gravity included a term he called the "cosmological constant," which he represented by the capital Greek letter lambda: Λ. A mathematically permitted but optional term, the cosmological constant allowed him to represent a static universe.

Back then, the idea that our universe would be doing anything at all, other than simply existing, was beyond anyone's imagination. So lambda's sole job was to oppose gravity within Einstein's model, keeping the universe in balance, resisting the natural tendency for gravity to pull the whole universe into one

[†] R. Israel, E. Ruckhaber, R. Weinmann, et al., *Hundert Autoren Gegen Einstein* (Leipzig: R. Voigtlanders Verlag, 1931).

giant mass. In this way, Einstein invented a universe that neither expands nor contracts, consistent with everybody's expectations at the time.

The Russian physicist Alexander Friedmann would subsequently show mathematically that Einstein's universe, though balanced, was in an unstable state. Like a ball resting on the top of a hill, awaiting the slightest provocation to roll down in one direction or another, or like a pencil balanced on its sharpened point, Einstein's universe was precariously perched between a state of expansion and total collapse. Moreover, Einstein's theory was new, and just because you give something a name does not make it real—Einstein knew that lambda, as a negative gravity force of nature, had no known counterpart in the physical universe.

*

Einstein's general theory of relativity radically departed from all previous thinking about gravitational attraction. Instead of settling for Sir Isaac Newton's view of gravity as spooky action-at-a-distance (a conclusion that made Newton himself uncomfortable), GR regards gravity as the response of a mass to the local curvature of space and time caused by some other mass or field of energy. In other words, concentrations of mass cause distortions—dimples, really—in the fabric of space and time. These distortions guide the moving masses along straight-line geodesics,[†] though they look to us like the curved trajectories we call orbits. The twentieth-century American theoretical physicist John Archibald Wheeler said it best, summing up Einstein's concept as, "Matter tells space how to curve; space tells matter how to move."[‡]

At the end of the day, general relativity described two kinds of gravity. One is the familiar kind, like the attraction between Earth

---

[†] "Geodesic" is a needlessly fancy word for the shortest distance between two points along a curved surface—extended, in this case, to be the shortest distance between two points in the curved four-dimensional fabric of space-time.

[‡] In graduate school I took John Wheeler's class on general relativity (where I met my wife) and he said this often.

and a ball thrown into the air, or between the Sun and the planets. It also predicted another variety—a mysterious, anti-gravity pressure associated with the vacuum of space-time itself. Lambda preserved what Einstein and every other physicist of his day had strongly presumed to be true: the status quo of a static universe—an unstable static universe. To invoke an unstable condition as the natural state of a physical system violates scientific credo. You cannot assert that the entire universe is a special case that happens to be balanced forever and ever. Nothing ever seen, measured, or imagined has behaved this way in the history of science, which makes for powerful precedent.

Thirteen years later, in 1929, the American astrophysicist Edwin P. Hubble discovered that the universe is not static. He had found and assembled convincing evidence that the more distant a galaxy, the faster the galaxy recedes from the Milky Way. In other words, the universe is expanding. Now, embarrassed by the cosmological constant, which corresponded to no known force of nature, and by the lost opportunity to have predicted the expanding universe himself, Einstein discarded lambda entirely, calling it his life's "greatest blunder." By yanking lambda from the equation he presumed its value to be zero, such as in this example: Assume $A = B + C$. If you learn later that $A = 10$ and $B = 10$, then A still equals B plus C, except in that case C equals 0 and is rendered unnecessary in the equation.

But that wasn't the end of the story. Off and on over the decades, theorists would extract lambda from the crypt, imagining what their ideas would look like in a universe that had a cosmological constant. Sixty-nine years later, in 1998, science exhumed lambda one last time. Early that year, remarkable announcements were made by two competing teams of astrophysicists: one led by Saul Perlmutter of Lawrence Berkeley National Laboratory in Berkeley, California, and the other co-led by Brian Schmidt of Mount Stromlo and Siding Spring observatories in Canberra, Australia, and Adam Riess of the Johns Hopkins University in Baltimore, Maryland. Dozens of the most distant supernovas ever observed appeared noticeably dimmer than expected, given the well-documented behavior of this species of exploding star. Reconciliation required that either those distant supernovas behaved unlike their nearer brethren, or they were as much as fifteen percent farther away than where the prevailing cosmological models had placed them. The only known thing that "naturally" accounts for this acceleration is Einstein's lambda, the cosmological constant. When astrophysicists dusted it off and put it back into Einstein's original equations for general relativity, the known state of the universe matched the state of Einstein's equations.

*

The supernovas used in Perlmutter's and Schmidt's studies are worth their weight in fusionable nuclei. Within certain limits, each of those stars explodes the same way, igniting the same amount of fuel, releasing the same titanic amount of energy in the same amount of time, thereby reaching the same peak luminosity. Thus they serve as a kind of yardstick, or "standard candle," for calculating cosmic distances to the galaxies in which they explode, out to the farthest reaches of the universe.

Standard candles simplify calculations immensely: since the supernovas all have the same wattage, the dim ones are far away and the bright ones are close by. After measuring their brightness (a simple task), you can tell exactly how far they are from you and from one another. If the luminosities of the supernovas were all different, you could not use brightness alone to tell how far away one was in comparison with another. A dim one could be either a high-wattage bulb far away or a low-wattage bulb close up.

All fine. But there's a second way to measure the distance to galaxies: their speed of recession from our Milky Way—recession that's part and parcel of the overall cosmic expansion. As Hubble was the first to show, the expanding universe makes distant objects race away from us faster than nearby ones. So, by measuring a galaxy's speed of recession (another simple task), one can deduce a galaxy's distance.

If those two well-tested methods give different distances for the same object, something must be wrong. Either the supernovas are bad standard candles, or our model for the rate of cosmic expansion as measured by galaxy speeds is wrong.

Well, something *was* wrong. It turned out that the supernovas were splendid standard candles, surviving the careful scrutiny of many skeptical investigators, and so astrophysicists were left with a universe that had expanded faster than we thought, placing galaxies farther away than their recession speed would have otherwise indicated. And there was no easy way to explain the extra expansion without invoking lambda, Einstein's cosmological constant.

Here was the first direct evidence that a repulsive force permeated the universe, opposing gravity, which is how and why the cosmological constant rose from the dead. Lambda suddenly acquired a physical reality that needed a name, and so "dark energy" took center stage in the cosmic drama, suitably capturing both the mystery and our associated ignorance of its cause. Perlmutter, Schmidt, and Reiss justifiably shared the 2011 Nobel Prize in physics for this discovery.

The most accurate measurements to date reveal dark energy as the most prominent thing in town, currently responsible for 68 percent of all the mass-energy in the universe; dark matter comprises 27 percent, with regular matter comprising a mere 5 percent.

✳

The shape of our four-dimensional universe comes from the relationship between the amount of matter and energy that lives in the cosmos and the rate at which the cosmos is expanding. A convenient mathematical measure of this is omega: Ω, yet another capital Greek letter with a firm grip on the cosmos.

If you take the matter-energy density of the universe and divide it by the matter-energy density required to just barely halt the expansion (known as the "critical" density), you get omega.

Since both mass and energy cause space-time to warp, or curve, omega tells us the shape of the cosmos. If omega is less than one, the actual mass-energy falls below the critical value, and the universe expands forever in every direction for all of time, taking on the shape of a saddle, in which initially parallel lines diverge. If omega equals one, the universe expands forever, but only barely so. In that case the shape is flat, preserving all the geometric rules we learned in high school about parallel lines. If omega exceeds one, parallel lines converge, and the universe curves back on itself, ultimately recollapsing into the fireball whence it came.

At no time since Hubble discovered the expanding universe has any team of observers ever reliably measured omega to be anywhere close to one. Adding up all the mass and energy their telescopes could see, and even extrapolating beyond these limits, dark matter included, the biggest values from the best observations topped out at about $\Omega = 0.3$. As far as observers were concerned, the universe was "open" for business, riding a one-way saddle into the future.

Meanwhile, beginning in 1979, the American physicist Alan H. Guth of the Massachusetts Institute of Technology, and others, advanced an adjustment to the big bang theory that cleared up some nagging problems with getting a universe to be as smoothly filled with matter and energy as ours is known to be. A fundamental by-product of this update to the big bang was that it drives omega toward one. Not toward a half. Not toward two. Not toward a million. Toward one.

Hardly a theorist in the world had a problem with that requirement, because it helped get the big bang to account for the global properties of the known universe. There was, however, another little problem: the update predicted three times as much mass-energy as observers could find. Undeterred, the theorists said the observers just weren't looking hard enough.

At the end of the tallies, visible matter alone could account for no more than 5 percent of the critical density. How about the mysterious dark matter? They added that, too. Nobody knew what it was, and we still don't know what it is, but surely it contributed to the totals. From there we get five or six times as much dark matter as visible matter. But that's still way too little. Observers were at a loss, and the theorists answered, "Keep looking."

Both camps were sure the other was wrong—

until the discovery of dark energy. That single component, when added to the ordinary matter and the ordinary energy and dark matter, raised the mass-energy density of the universe to the critical level. Simultaneously satisfying both the observers and the theorists.

For the first time, the theorists and observers kissed and made up. Both, in their own way, were correct. Omega does equal one, just as the theorists demanded of the universe, even though you can't get there by adding up all the matter—dark or otherwise—as they had naively presumed. There's no more matter running around the cosmos today than had ever been estimated by the observers.

Nobody had foreseen the dominating presence of cosmic dark energy, nor had anybody imagined it as the great reconciler of differences.

*

So what is the stuff? Nobody knows. The closest anybody has come is to presume dark energy is a quantum effect—where the vacuum of space, instead of being empty, actually seethes with particles and their antimatter counterparts. They pop in and out of existence in pairs, and don't last long enough to be measured. Their transient existence is captured in their moniker: virtual particles. The remarkable legacy of quantum physics—the science of the small—demands that we give this idea serious attention. Each pair of virtual particles exerts a little bit of outward pressure as it ever so briefly elbows its way into space.

Unfortunately, when you estimate the amount of repulsive "vacuum pressure" that arises from the abbreviated lives of virtual particles, the result is more than 10^{120} times larger than the experimentally determined value of the cosmological constant. This is a stupidly large factor, leading to the biggest mismatch between theory and observation in the history of science.

Yes, we're clueless. But it's not abject cluelessness. Dark energy is not adrift, with nary a theory to anchor it. Dark energy inhabits one of the safest harbors we can imagine: Einstein's equations of general relativity. It's the cosmological constant. It's lambda. Whatever dark energy turns out to be, we already know how to measure it and how to calculate its effects on the past, present, and future of the cosmos.

Without a doubt, Einstein's greatest blunder was having declared that lambda was his greatest blunder.

＊

And the hunt is on. Now that we know dark energy is real, teams of astrophysicists have begun ambitious programs to measure distances and the growth of structure in the universe using ground-based and spaceborne telescopes. These observations will test the detailed influence of dark energy on the expansion history of the universe, and will surely keep theorists busy. They desperately need to atone for how embarrassing their calculation of dark energy turned out to be.

Do we need an alternative to GR? Does the marriage of GR and quantum mechanics require an overhaul? Or is there some theory of dark energy that awaits discovery by a clever person yet to be born?

A remarkable feature of lambda and the

accelerating universe is that the repulsive force arises from within the vacuum, not from anything material. As the vacuum grows, the density of matter and (familiar) energy within the universe diminishes, and the greater becomes lambda's relative influence on the cosmic state of affairs. With greater repulsive pressure comes more vacuum, and with more vacuum comes greater repulsive pressure, forcing an endless and exponential acceleration of the cosmic expansion.

As a consequence, anything not gravitationally bound to the neighborhood of the Milky Way galaxy will recede at ever-increasing speed, as part of the accelerating expansion of the fabric of space-time. Distant galaxies now visible in the night sky will ultimately disappear beyond an unreachable horizon, receding from us faster than the speed of light. A feat allowed, not because they're moving through space at such speeds, but because the fabric of the universe itself carries them at such speeds. No law of physics prevents this.

In a trillion or so years, anyone alive in our own galaxy may know nothing of other galaxies. Our observable universe will merely comprise a system of nearby, long-lived stars within the Milky Way. And beyond this starry night will lie an endless void—darkness in the face of the deep.

Dark energy, a fundamental property of the cosmos, will, in the end, undermine the ability of future generations to comprehend the universe they've been dealt. Unless contemporary astrophysicists across the galaxy keep remarkable records and bury an awesome, trillion-year time capsule, postapocaplyptic scientists will know nothing of galaxies—the principal form of organization for matter in our cosmos—and will thus be denied access to key pages from the cosmic drama that is our universe.

Behold my recurring nightmare: Are we, too, missing some basic pieces of the universe that once were? What part of the cosmic history book has been marked "access denied"? What remains absent from our theories and equations that ought to be there, leaving us groping for answers we may never find?

7.

The Cosmos
on the Table

Trivial questions sometimes require deep and expansive knowledge of the cosmos just to answer them. In middle school chemistry class, I asked my teacher where the elements on the Periodic Table come from. He replied, Earth's crust. I'll grant him that. It's surely where the supply lab gets them. But how did Earth's crust acquire them? The answer must be astronomical. But in this case, do you actually need to know the origin and evolution of the universe to answer the question?

Yes, you do.

Only three of the naturally occurring elements were manufactured in the big bang. The rest were forged in the high-temperature hearts and explosive remains of dying stars, enabling subsequent generations of star systems to incorporate this enrichment, forming planets and, in our case, people.

For many, the Periodic Table of Chemical Elements is a forgotten oddity—a chart of boxes filled with mysterious, cryptic letters last encountered on the wall of high school chemistry class. As the organizing principle for the chemical behavior of all known and yet-to-be-discovered elements in the universe, the table instead ought to be a cultural icon, a testimony to the enterprise of science as an international human adventure conducted in laboratories, particle accelerators, and on the frontier of the cosmos itself.

Yet every now and then, even a scientist can't help thinking of the Periodic Table as a zoo of one-of-a-kind animals conceived by Dr. Seuss. How else could we believe that sodium is a poisonous, reactive metal that you can cut with a butter knife, while pure chlorine is a smelly, deadly gas, yet when added together they make sodium chloride, a harmless, biologically essential compound better known as table salt? Or how about hydrogen and oxygen? One is an explosive gas, and the other promotes violent combustion, yet the two combined make liquid water, which puts out fires.

Amid these chemical confabulations we find elements significant to the cosmos, allowing me to offer the Periodic Table as viewed through the lens of an astrophysicist.

*

With only one proton in its nucleus, hydrogen is the lightest and simplest element, made entirely during the big bang. Out of the ninety-four naturally occurring elements, hydrogen lays claim to more than two-thirds of all the atoms in the human body, and more than ninety percent of all atoms in the cosmos, on all scales, right on down to the solar system. Hydrogen in the core of the massive planet Jupiter is under so much pressure that it behaves more like a conductive metal than a gas, creating the strongest magnetic field among the planets. The English chemist Henry Cavendish discovered hydrogen in 1766 during his experiments with H_2O (*hydro-genes* is Greek for "water-forming"), but he is best known among astrophysicists as the first to calculate Earth's mass after having measured an accurate value for big G in Newton's famous equation for gravity.

Every second of every day, 4.5 billion tons of fast-moving hydrogen nuclei are turned into energy as they slam together to make helium within the fifteen-million-degree core of the Sun.

*

Helium is widely recognized as an over-the-counter, low-density gas that, when inhaled, temporarily increases the vibrational frequency of your windpipe and larynx, making you sound like Mickey Mouse. Helium is the second simplest and second most abundant element in the universe. Although a distant second to hydrogen in abundance, there's four times more of it than all other elements in the universe combined. One of the pillars of big bang cosmology is the prediction that in every region of the cosmos, no less than about ten percent of all atoms are helium, manufactured in that percentage across the well-mixed primeval fireball that was the birth of our universe. Since the thermonuclear fusion of hydrogen within stars gives you helium, some regions of the cosmos could easily accumulate more than their ten percent share of helium, but, as predicted, no one has ever found a region of the galaxy with less.

Some thirty years before it was discovered and isolated on Earth, astronomers detected helium in the spectrum of the Sun's corona during the total eclipse of 1868. As noted earlier, the name helium was duly derived from Helios, the Greek sun god. And with 92 percent of hydrogen's buoyancy in air, but without its explosive characteristics, helium is the gas of choice for the outsized balloon characters of the Macy's Thanksgiving Day parade, making the department store second only to the U.S. military as the nation's top user of the element.

*

Lithium is the third simplest element in the universe, with three protons in its nucleus. Like hydrogen and helium, lithium was made in the big bang, but unlike helium, which can be manufactured in stellar cores, lithium is destroyed by every known nuclear reaction. Another prediction of big bang cosmology is that we can expect no more than one percent of the atoms in any region of the universe to be lithium. No one has yet found a galaxy with more lithium than this upper limit supplied by the big bang. The combination of helium's upper limit and lithium's lower limit gives a potent dual-constraint on tests for big bang cosmology.

*

The element carbon can be found in more kinds of molecules than the sum of all other kinds of molecules combined. Given the abundance of carbon in the cosmos— forged in the cores of stars, churned up to their surfaces, and released copiously into the galaxy—a better element does not exist on which to base the chemistry and diversity of life. Just edging out carbon in abundance rank, oxygen is common, too, forged and released in the remains of exploded stars.

Both oxygen and carbon are major ingredients of life as we know it.

But what about life as we don't know it? How about life based on the element silicon? Silicon sits directly below carbon on the Periodic Table, which means, in principle, it can create the same portfolio of molecules that carbon does. In the end, we expect carbon to win because it's ten times more abundant than silicon in the cosmos. But that doesn't stop science fiction writers, who keep exobiologists on their toes, wondering what the first truly alien, silicon-based life forms would be like.

In addition to being an active ingredient in table salt, at the moment, sodium is the most common glowing gas in municipal street lamps across the nation. They "burn" brighter and longer than incandescent bulbs, although they may all soon be replaced by LEDs, which are even brighter at a given wattage, and cheaper. Two varieties of sodium lamps are common: high-pressure lamps, which look yellow-white, and the rarer low-pressure lamps, which look orange. Turns out, while all light pollution is bad for astrophysics, the low-pressure sodium lamps

are least bad because their contamination can be
easily subtracted from telescope data. In a model
of cooperation, the entire city of Tucson, Arizona,
the nearest large municipality to the Kitt Peak
National Observatory, has, by agreement with
the local astrophysicists, converted all its streetlights to low-pressure sodium lamps.

*

Aluminum occupies nearly ten percent of
Earth's crust yet was unknown to the ancients
and unfamiliar to our great-grandparents. The
element was not isolated and identified until
1827 and did not enter common household use
until the late 1960s, when tin cans and tin foil
yielded to aluminum cans and, of course, aluminum foil. (I'd bet most old people you know
still call the stuff tin foil.) Polished aluminum
makes a near-perfect reflector of visible light
and is the coating of choice for nearly all telescope mirrors today.

Titanium is 1.7 times denser than aluminum,
but it's more than twice as strong. So titanium,
the ninth most abundant element in Earth's
crust, has become a modern darling for many

applications, such as military aircraft components and prosthetics that require a light, strong
metal for their tasks.

In most cosmic places, the number of oxygen
atoms exceeds that of carbon. After every carbon atom has latched onto the available oxygen
atoms (forming carbon monoxide or carbon dioxide), the leftover oxygen bonds with other things,
like titanium. The spectra of red stars are riddled
with features traceable to titanium oxide, which
itself is no stranger to stars on Earth: star sap-
phires and rubies owe their radiant asterisms to
titanium oxide impurities in their crystal lattice.
Furthermore, the white paint used for telescope
domes features titanium oxide, which happens to
be highly reflective in the infrared part of the
spectrum, greatly reducing the heat accumulated
from sunlight in the air surrounding the telescope. At nightfall, with the dome open, the air
temperature near the telescope rapidly equals the
temperature of the nighttime air, allowing light
from stars and other cosmic objects to be sharp
and clear. And, while not directly named for a
cosmic object, titanium derives from the Titans of
Greek mythology; Titan is Saturn's largest moon.

*

By many measures, iron ranks as the most important element in the universe. Massive stars manufacture elements in their core, in sequence from helium to carbon to oxygen to nitrogen, and so forth, all the way up the Periodic Table to iron. With twenty-six protons and at least as many neutrons in its nucleus, iron's odd distinction comes from having the least total energy per nuclear particle of any element. This means something quite simple: if you split iron atoms via fission, they will absorb energy. And if you combine iron atoms via fusion, they will also absorb energy. Stars, however, are in the business of making energy. As high-mass stars manufacture and accumulate iron in their cores, they are nearing death. Without a fertile source of energy, the star collapses under its own weight and instantly rebounds in a stupendous supernova explosion, outshining a billion suns for more than a week.

*

The soft metal gallium has such a low melting point that, like cocoa butter, it will liquefy on contact with your hand. Apart from this parlor demo, gallium is not interesting to astrophysicists, except as one of the ingredients in the gallium chloride experiments used to detect elusive neutrinos from the Sun. A huge (100-ton) underground vat of liquid gallium chloride is monitored for any collisions between neutrinos and gallium nuclei, turning it into germanium. The encounter emits a spark of X-ray light that is measured every time a nucleus gets slammed. The long-standing solar neutrino problem, where fewer neutrinos were detected than predicted by solar theory, was solved using "telescopes" such as this.

*

Every form of the element technetium is radioactive. Not surprisingly, it's found nowhere on Earth except in particle accelerators, where we make it on demand. Technetium carries this distinction in its name, which derives from the Greek *technetos*, meaning "artificial." For reasons not yet fully understood, technetium lives in the atmospheres of a select subset of red stars. This alone would not be cause for alarm except that technetium has a half-life of a mere two million years, which is much, much shorter than the age and life expectancy of the stars in which it is found. In other words, the star cannot have been born with the stuff, for if it were, there would be none left by now. There is also no known mechanism to create technetium in a star's core *and* have it dredge itself up to the surface where it is observed, which has led to exotic theories that have yet to achieve consensus in the astrophysics community.

*

Along with osmium and platinum, iridium is one of the three heaviest (densest) elements on the Table—two cubic feet of it weighs as much as a Buick, which makes iridium one of the world's best paperweights, able to defy all known office fans. Iridium is also the world's most famous smoking gun. A thin layer of it can be found worldwide at the famous Cretaceous-Paleogene (K-Pg) boundary[†] in geological strata, dating from sixty-five million years ago. Not so coincidentally, that's when every

---

[†] For old-timers, this layer was formerly known as the Cretaceous-Tertiary (K-T) boundary.

land species larger than a carry-on suitcase went extinct, including the legendary dinosaurs. Iridium is rare on Earth's surface but relatively common in six-mile metallic asteroids, which, upon colliding with Earth, vaporize on impact, scattering their atoms across Earth's surface. So, whatever might have been your favorite theory for offing the dinosaurs, a killer asteroid the size of Mount Everest from outer space should be at the top of your list.

*

I don't know how Albert would have felt about this, but an unknown element was discovered in the debris of the first hydrogen bomb test in the Eniwetok atoll in the South Pacific, on November 1, 1952, and was named einsteinium in his honor. I might have named it armageddium instead.

Meanwhile, ten entries in the Periodic Table get their names from objects that orbit the Sun:

Phosphorus comes from the Greek for "light-bearing," and was the ancient name for the planet Venus when it appeared before sunrise in the dawn sky.

Selenium comes from *selene*, which is Greek

for the Moon, named so because in ores, it was always associated with the element tellurium, which had already been named for Earth, from the Latin *tellus*.

On January 1, 1801, the Italian astronomer Giuseppe Piazzi discovered a new planet orbiting the Sun in the suspiciously large gap between Mars and Jupiter. Keeping with the tradition of naming planets after Roman gods, the object was named Ceres, after the goddess of harvest. Ceres is, of course, the root of the word "cereal." At the time, there was sufficient excitement in the scientific community for the first element discovered after this date to be named cerium in its honor. Two years later, another planet was discovered, orbiting the Sun in the same gap as Ceres. This one was named Pallas, for the Roman goddess of wisdom, and, like cerium before it, the first element discovered thereafter was named palladium in its honor. The naming party would end a few decades later. After dozens more of these planets were discovered sharing the same orbital zone, closer analysis revealed that these objects were much, much smaller than the smallest known planets. A new swath of real estate had been discovered in the solar system, populated by small, craggy chunks of rock and metal. Ceres and Pallas were not planets; they are asteroids, and they live in the asteroid belt, now known to contain hundreds of thousands of objects—somewhat more than the number of elements in the Periodic Table.

The metal mercury, liquid and runny at room temperature, and the planet Mercury, the fastest of all planets in the solar system, are both named for the speedy Roman messenger god of the same name.

Thorium is named for Thor, the hunky, lightning bolt–wielding Scandinavian god, who corresponds with lightning bolt–wielding Jupiter in Roman mythology. And by Jove, Hubble Space Telescope images of Jupiter's polar regions reveal extensive electrical discharges deep within its turbulent cloud layers.

Alas, Saturn, my favorite planet,[†] has no element named for it, but Uranus, Neptune, and Pluto are famously represented. The element

[†] Actually, Earth is my favorite planet. Then Saturn.

uranium was discovered in 1789 and named in honor of the planet discovered by William Herschel just eight years earlier. All isotopes of uranium are unstable, spontaneously decaying to lighter elements, a process accompanied by the release of energy. The first atomic bomb ever used in warfare had uranium as its active ingredient, and was dropped by the United States, incinerating the Japanese city of Hiroshima on August 6, 1945. With ninety-two protons packed in its nucleus, uranium is widely described as the "largest" naturally occurring element, although trace amounts of larger elements can be found naturally where uranium ore is mined.

If Uranus deserved an element named in its honor, then so did Neptune. Unlike uranium, however, which was discovered shortly after the planet, neptunium was discovered in 1940 in the Berkeley cyclotron, a full ninety-seven years after the German astronomer John Galle found Neptune in a spot on the sky predicted by the French mathematician Joseph Le Verrier after studying Uranus's odd orbital behavior. Just as Neptune comes right after Uranus in the solar system, so too does neptunium come right after uranium in the Periodic Table of elements.

The Berkeley cyclotron discovered (or manufactured?) many elements not found in nature,
including plutonium, which directly follows
neptunium in the table and was named for
Pluto, which Clyde Tombaugh discovered at
Arizona's Lowell Observatory in 1930. Just as
with the discovery of Ceres 129 years earlier,
excitement prevailed. Pluto was the first planet
discovered by an American and, in the absence
of better data, was widely regarded as an object
of commensurate size and mass to Earth, if
not Uranus or Neptune. As our attempts to
measure Pluto's size became more and more
refined, Pluto kept getting smaller and smaller.
Our knowledge of Pluto's dimensions did not
stabilize until the late 1980s. We now know
that cold, icy Pluto is by far the smallest of the
nine, with the diminutive distinction of being
littler than the solar system's six largest moons.
And like the asteroids, hundreds more objects
were later discovered in the outer solar system
with orbits similar to that of Pluto, signaling
the end of Pluto's tenure as a planet, and the
revelation of a heretofore undocumented reservoir of small icy bodies called the Kuiper
belt of comets, to which Pluto belongs. In this

regard, one could argue that Ceres, Pallas, and Pluto slipped into the Periodic Table under false pretenses.

Unstable weapons-grade plutonium was the active ingredient in the atomic bomb that the United States exploded over the Japanese city of Nagasaki, just three days after Hiroshima, bringing a swift end to World War II. Small quantities of non-weapons-grade radioactive plutonium can be used to power radioisotope thermoelectric generators (sensibly abbreviated as RTGs) for spacecraft that travel to the outer solar system, where the intensity of sunlight has diminished below the level usable by solar panels. One pound of plutonium will generate ten million kilowatt-hours of heat energy, which is enough to power an incandescent lightbulb for eleven thousand years, or a human being for just as long if we ran on nuclear fuel instead of grocery-store food.

*

So ends our cosmic journey through the Periodic Table of Chemical Elements, right to the edge of the solar system, and beyond. For reasons I have yet to understand, many people don't like chemicals, which might explain the perennial movement to rid foods of them. Perhaps sesquipedalian chemical names just sound dangerous. But in that case we should blame the chemists, and not the chemicals themselves. Personally, I am quite comfortable with chemicals, anywhere in the universe. My favorite stars, as well as my best friends, are all made of them.

8.

On Being Round

Apart from crystals and broken rocks, not much else in the cosmos naturally comes with sharp angles. While many objects have peculiar shapes, the list of round things is practically endless and ranges from simple soap bubbles to the entire observable universe. Of all shapes, spheres are favored by the action of simple physical laws. So prevalent is this tendency that often we assume something is spherical in a mental experiment just to glean basic insight even when we know that the object is decidedly non-spherical. In short, if you do not understand the spherical case, then you cannot claim to understand the basic physics of the object.

Spheres in nature are made by forces, such as surface tension, that want to make objects smaller in all directions. The surface tension of the liquid that makes a soap bubble squeezes air in all directions. It will, within moments of being formed, enclose the volume of air using the least possible surface area. This makes the strongest possible bubble because the soapy film will not have to be spread any thinner than is absolutely necessary. Using freshman-level calculus you can show that the one and only shape that has the smallest surface area for an enclosed volume is a perfect sphere. In fact, billions of dollars could be saved annually on packaging materials if all shipping boxes and all packages of food in the supermarket were spheres. For example, the contents of a super-jumbo box of Cheerios would fit easily into a spherical carton with a four-and-a-half-inch radius. But practical matters prevail— nobody wants to chase packaged food down the aisle after it rolls off the shelves.

On Earth, one way to make ball bearings is to machine them, or drop molten metal in pre-measured amounts into the top of a long shaft. The blob will typically undulate until it settles into the shape of a sphere, but it needs sufficient time to harden before hitting the bottom. On orbiting space stations, where everything is weightless, you gently squirt out precise quantities of molten metal and you have all the time you need—the beads just float there while they cool, until they harden as perfect spheres, with surface tension doing all the work for you.

*

For large cosmic objects, energy and gravity conspire to turn objects into spheres. Gravity is the force that serves to collapse matter in all directions, but gravity does not always win—chemical bonds of solid objects are strong. The Himalayas grew against the force of Earth's gravity because of the resilience of crustal rock. But before you get excited about Earth's mighty mountains, you should know that the spread in height from the deepest undersea trenches to the tallest mountains is about a dozen miles, yet Earth's diameter is nearly eight thousand miles. So, contrary to what it looks like to teeny humans crawling on its surface, Earth, as a cosmic object, is remarkably smooth. If you had a super-duper, jumbo-gigantic finger, and you dragged it across Earth's surface (oceans and all), Earth would feel as smooth as a cue ball. Expensive globes that portray raised portions of Earth's landmasses to indicate mountain ranges are gross exaggerations of reality. This is why, in spite of Earth's mountains and valleys, as well as being slightly flattened from pole to pole, when viewed from space, Earth is indistinguishable from a perfect sphere.

Earth's mountains are also puny when compared with some other mountains in the solar system. The largest on Mars, Olympus Mons, is 65,000 feet tall and nearly 300 miles wide at its base. It makes Alaska's Mount McKinley look like a molehill. The cosmic mountain-building recipe is simple: the weaker the gravity on the surface of an object, the higher its mountains can reach. Mount Everest is about as tall as a mountain on Earth can grow before the lower rock layers succumb to their own plasticity under the mountain's weight.

If a solid object has a low enough surface gravity, the chemical bonds in its rocks will resist the force of their own weight. When this happens, almost any shape is possible. Two famous celestial non-spheres are Phobos and Deimos, the Idaho potato–shaped moons of Mars. On thirteen-mile-long Phobos, the bigger of the two moons, a 150-pound person would weigh a mere four ounces.

In space, surface tension always forces a small blob of liquid to form a sphere. Whenever you see a small solid object that is suspiciously spherical, you can assume it formed in a molten state. If the blob has very high mass, then it could be composed of almost anything and gravity will ensure that it forms a sphere.

Big and massive blobs of gas in the galaxy can coalesce to form near-perfect, gaseous spheres called stars. But if a star finds itself orbiting too close to another object whose gravity is significant, the spherical shape can be distorted as its material gets stripped away. By "too close," I mean too close to the object's Roche lobe—named for the mid-nineteenth-century mathematician Édouard Roche, who made detailed studies of gravity fields in the vicinity of double stars. The Roche lobe is a theoretical, dumbbell-shaped, bulbous, double envelope that surrounds any two objects in mutual orbit.

If gaseous material from one object passes out
of its own envelope, then the material will fall
toward the second object. This occurrence is
common among binary stars when one of them
swells to become a red giant and overfills its
Roche lobe. The red giant distorts into a dis-
tinctly non-spherical shape that resembles an
elongated Hershey's kiss. Moreover, every now
and then, one of the two stars is a black hole,
whose location is rendered visible by the flay-
ing of its binary companion. The spiraling gas,
after having passed from the giant across its
Roche lobe, heats to extreme temperatures and
is rendered aglow before descending out of sight
into the black hole itself.

*

The stars of the Milky Way galaxy trace a
big, flat circle. With a diameter-to-thickness
ratio of one thousand to one, our galaxy is flatter than the flattest flapjacks ever made. In
fact, its proportions are better represented by
a crépe or a tortilla. No, the Milky Way's disk
is not a sphere, but it probably began as one.
We can understand the flatness by assuming
the galaxy was once a big, spherical, slowly

rotating ball of collapsing gas. During the collapse, the ball spun faster and faster, just as spinning figure skaters do when they draw their arms inward to increase their rotation rate. The galaxy naturally flattened pole-to-pole while the increasing centrifugal forces in the middle prevented collapse at midplane. Yes, if the Pillsbury Doughboy were a figure skater, then fast spins would be a high-risk activity.

Any stars that happened to be formed within the Milky Way cloud before the collapse maintained large, plunging orbits. The remaining gas, which easily sticks to itself, like a mid-air collision of two hot marshmallows, got pinned at the mid-plane and is responsible for all subsequent generations of stars, including the Sun. The current Milky Way, which is neither collapsing nor expanding, is a gravitationally mature system where one can think of the orbiting stars above and below the disk as the skeletal remains of the original spherical gas cloud.

This general flattening of objects that rotate is why Earth's pole-to-pole diameter is smaller than its diameter at the equator. Not by much: three-tenths of one percent—about twenty-six miles. But Earth is small, mostly solid, and doesn't rotate all that fast. At twenty-four hours per day, Earth carries anything on its equator at a mere 1,000 miles per hour. Consider the jumbo, fast-rotating, gaseous planet Saturn. Completing a day in just ten and a half hours, its equator revolves at 22,000 miles per hour and its pole-to-pole dimension is a full ten percent flatter than its middle, a difference noticeable even through a small amateur telescope. Flattened spheres are more generally called oblate spheroids, while spheres that are elongated pole-to-pole are called prolate. In everyday life, hamburgers and hot dogs make excellent (although somewhat extreme) examples of each shape. I don't know about you, but the planet Saturn pops into my mind with every bite of a hamburger I take.

*

We use the effect of centrifugal forces on matter to offer insight into the rotation rate of extreme cosmic objects. Consider pulsars.

With some rotating at upward of a thousand revolutions per second, we know that they cannot be made of household ingredients, or they would spin themselves apart. In fact, if a pulsar rotated any faster, say 4,500 revolutions per second, its equator would be moving at the speed of light, which tells you that this material is unlike any other. To picture a pulsar, imagine the mass of the Sun packed into a ball the size of Manhattan. If that's hard to do, then maybe it's easier if you imagine stuffing about a hundred million elephants into a Chapstick casing. To reach this density, you must compress all the empty space that atoms enjoy around their nucleus and among their orbiting electrons. Doing so will crush nearly all (negatively charged) electrons into (positively charged) protons, creating a ball of (neutrally charged) neutrons with a crazy-high surface gravity. Under such conditions, a neutron star's mountain range needn't be any taller than the thickness of a sheet of paper for you to exert more energy climbing it than a rock climber on Earth would exert ascending a three-thousand-mile-high cliff. In short, where gravity is high, the high places tend to fall, filling in the low places—a phenomenon that sounds almost biblical, in preparing the way for the Lord: "Every valley shall be raised up, every mountain and hill made low; the rough ground shall become level, the rugged places a plain" (Isaiah 40:4). That's a recipe for a sphere if there ever was one. For all these reasons, we expect pulsars to be the most perfectly shaped spheres in the universe.

*

For rich clusters of galaxies, the overall shape can offer deep astrophysical insight. Some are raggedy. Others are stretched thin in filaments. Yet others form vast sheets. None of these have settled into a stable—spherical—gravitational shape. Some are so extended that the fourteen-billion-year age of the universe is insufficient time for their constituent galaxies to make one crossing of the cluster. We conclude that the cluster was born that way because the mutual gravitational encounters between and among galaxies have had insufficient time to influence the cluster's shape.

But other systems, such as the beautiful
Coma cluster of galaxies, which we met in
our chapter on dark matter, tell us immediately that gravity has shaped the cluster into
a sphere. As a consequence, you are as likely
to find a galaxy moving in one direction as
in any other. Whenever this is true, the cluster cannot be rotating all that fast; otherwise,
we would see some flattening, as we do in our
own Milky Way.

The Coma cluster, once again like the
Milky Way, is also gravitationally mature. In
astrophysical vernacular, such systems are said
to be "relaxed," which means many things,
including the fortuitous fact that the average
velocity of galaxies in the cluster serves as an
excellent indicator of the total mass, whether
or not the total mass of the system is supplied
by the objects used to get the average velocity. It's for these reasons that gravitationally
relaxed systems make excellent probes of nonluminous "dark" matter. Allow me to make
an even stronger statement: were it not for
relaxed systems, the ubiquity of dark matter
may have remained undiscovered to this day.

*

The sphere to end all spheres—the largest and most perfect of them all—is the entire observable universe. In every direction we look, galaxies recede from us at speeds proportional to their distance. As we saw in the first few chapters, this is the famous signature of an expanding universe, discovered by Edwin Hubble in 1929. When you combine Einstein's relativity and the velocity of light and the expanding universe and the spatial dilution of mass and energy as a consequence of that expansion, there is a distance in every direction from us where the recession velocity for a galaxy equals the speed of light. At this distance and beyond, light from all luminous objects loses all its energy before reaching us. The universe beyond this spherical "edge" is thus rendered invisible and, as far as we know, unknowable.

There's a variation of the ever-popular multiverse idea in which the multiple universes that comprise it are not separate universes entirely, but isolated, non-interacting pockets of space

within one continuous fabric of space-time—
like multiple ships at sea, far enough away from
one another so that their circular horizons do
not intersect. As far as any one ship is concerned (without further data), it's the only ship
on the ocean, yet they all share the same body
of water.

*

Spheres are indeed fertile theoretical tools
that help us gain insight into all manner of
astrophysical problems. But one should not
be a sphere-zealot. I am reminded of the
half-serious joke about how to increase milk
production on a farm: An expert in animal
husbandry might say, "Consider the role of the
cow's diet . . ." An engineer might say, "Consider the design of the milking machines . . ."
But it's the astrophysicist who says, "Consider
a spherical cow . . ."

9.

Invisible Light

And therefore as a stranger give it welcome.
There are more things in heaven and earth, Horatio,
Than are dreamt of in your philosophy

HAMLET, ACT 1, SCENE 5

Before 1800 the word "light," apart from its use as a verb and an adjective, referred just to visible light. But early that year the English astronomer William Herschel observed some warming that could only have been caused by a form of light invisible to the human eye. Already an accomplished observer, Herschel had discovered the planet Uranus in 1781 and was now exploring the relation between sunlight, color, and heat. He began by placing a prism in the path of a

sunbeam. Nothing new there. Sir Isaac Newton had done that back in the 1600s, leading
him to name the familiar seven colors of the
visible spectrum: red, orange, yellow, green,
blue, indigo, and violet. (Yes, the colors do
indeed spell Roy G. Biv.) But Herschel was
inquisitive enough to wonder what the temperature of each color might be. So he placed
thermometers in various regions of the rain-
bow and showed, as he suspected, that different colors registered different temperatures.[†]

Well-conducted experiments require a
"control"—a measurement where you expect
no effect at all, and which serves as a kind
of idiot-check on what you are measuring.
For example, if you wonder what effect beer
has on a tulip plant, then also nurture a second tulip plant, identical to the first, but give
it water instead. If both plants die—if you

[†] Not until the mid-1800s, when the physicist's spectrom-
eter was applied to astronomical problems, did the astrono-
mer become the astrophysicist. In 1895, the prestigious
Astrophysical Journal was founded, with the subtitle "An
International Review of Spectroscopy and Astronomical
Physics."

killed them both—then you can't blame the alcohol. That's the value of a control sample. Herschel knew this, and laid a thermometer outside of the spectrum, adjacent to the red, expecting to read no more than room temperature throughout the experiment. But that's not what happened. The temperature of his control thermometer rose even higher than in the red.

Herschel wrote:

[I] conclude, that the full red falls still short of the maximum of heat; which perhaps lies even a little beyond visible refraction In this case, radiant heat will at least partly, if not chiefly, consist, if I may be permitted the expression, of invisible light; that is to say, of rays coming from the sun, that have such a momentum as to be unfit for vision.[†]

Holy s#%t!

† William Herschel, "Experiments on Solar and on the Terrestrial Rays that Occasion Heat," *Philosophical Transactions of the Royal Astronomical Society*, 1800, 17.

Herschel inadvertently discovered "infra" red light, a brand-new part of the spectrum found just "below" red, reported in the first of his four papers on the subject.

Herschel's revelation was the astronomical equivalent of Antonie van Leeuwenhoek's discovery of "many very little living animalcules, very prettily a-moving"[†] in the smallest drop of lake water. Leeuwenhoek discovered single-celled organisms—a biological universe. Herschel discovered a new band of light. Both hiding in plain sight.

Other investigators immediately took up where Herschel left off. In 1801 the German physicist and pharmacist Johann Wilhelm Ritter found yet another band of invisible light. But instead of a thermometer, Ritter placed a little pile of light-sensitive silver chloride in each visible color as well as in the dark area next to the violet end of the spectrum. Sure enough, the pile in the unlit patch darkened more than the pile in the violet patch.

[†] Antonie van Leeuwenhoek, letter to the Royal Society of London, October 10, 1676.

What's beyond violet? "Ultra" violet, better known today as UV.

Filling out the entire electromagnetic spectrum, in order of low-energy and low-frequency to high-energy and high-frequency, we have: radio waves, microwaves, infrared, ROYGBIV, ultraviolet, X-rays, and gamma rays. Modern civilization has deftly exploited each of these bands for countless household and industrial applications, making them familiar to us all.

*

After the discovery of UV and IR, sky-watching didn't change overnight. The first telescope designed to detect invisible parts of the electromagnetic spectrum wouldn't be built for 130 years. That's well after radio waves, X-rays, and gamma rays had been discovered, and well after the German physicist Heinrich Hertz had shown that the only real difference among the various kinds of light is the frequency of the waves in each band. In fact, credit Hertz for recognizing that there is such a thing as an electromagnetic spectrum. In his honor, the unit of frequency—in waves per second—

for anything that vibrates, including sound, has duly been named the hertz.

Mysteriously, astrophysicists were a bit slow to make the connection between the newfound invisible bands of light and the idea of building a telescope that might see those bands from cosmic sources. Delays in detector technology surely mattered here. But hubris must take some of the blame: how could the universe possibly send us light that our marvelous eyes cannot see? For more than three centuries—from Galileo's day until Edwin Hubble's—building a telescope meant only one thing: making an instrument to catch visible light, enhancing our biologically endowed vision.

A telescope is merely a tool to augment our meager senses, enabling us to get better acquainted with faraway places. The bigger the telescope, the dimmer the objects it brings into view; the more perfectly shaped its mirrors, the sharper the image it makes; the more sensitive its detectors, the more efficient its observations. But in all cases, every bit of information a telescope delivers to the astrophysicist comes to Earth on a beam of light.

Celestial happenings, however, don't limit themselves to what's convenient for the human retina. Instead, they typically emit varying amounts of light simultaneously in multiple bands. So without telescopes and their detectors tuned across the entire spectrum, astrophysicists would remain blissfully ignorant of some mind-blowing stuff in the universe.

Take an exploding star—a supernova. It's a cosmically common and seriously high-energy event that generates prodigious quantities of X-rays. Sometimes, bursts of gamma rays and flashes of ultraviolet accompany the explosions, and there's never a shortage of visible light. Long after the explosive gases cool, the shock waves dissipate, and the visible light fades, the supernova "remnant" keeps on shining in the infrared, while pulsing in radio waves. That's where pulsars come from, the most reliable timekeepers in the universe.

Most stellar explosions take place in distant galaxies, but if a star were to blow up within the Milky Way, its death throes would be bright enough for everyone to see, even without a telescope. But nobody on Earth saw the invisible

X-rays or gamma rays from the last two super-nova spectaculars hosted by our galaxy—one in 1572 and another in 1604—yet their wondrous visible light was widely reported.

The range of wavelengths (or frequencies) that comprise each band of light strongly influences the design of the hardware used to detect it. That's why no single combination of telescope and detector can simultaneously see every feature of such explosions. But the way around that problem is simple: gather all observations of your object, perhaps obtained by colleagues, in multiple bands of light. Then assign visible colors to invisible bands of interest, creating one meta, multi-band image. That's precisely what Geordi from the television series *Star Trek: The Next Generation* sees. With that power of vision, you miss nothing.

Only after you identify the band of your astrophysical affections can you begin to think about the size of your mirror, the materials you'll need to make it, the shape and surface it must have, and the kind of detector you'll need. X-ray wavelengths, for example, are extremely short. So if you're accumulating them, your

mirror had better be super-smooth, lest imper-
fections in the surface distort them. But if
you're gathering long radio waves, your mirror
could be made of chicken wire that you've bent
with your hands, because the irregularities
in the wire would be much smaller than the
wavelengths you're after. Of course, you also
want plenty of detail—high resolution—so
your mirror should be as big as you can afford
to make it. In the end, your telescope must be
much, much wider than the wavelength of
light you aim to detect. And nowhere is this
need more evident than in the construction of
a radio telescope.

*

Radio telescopes, the earliest non-visible-
light telescopes ever built, are an amazing subspecies of observatory. The American engineer
Karl G. Jansky built the first successful one
between 1929 and 1930. It looked a bit like the
moving sprinkler system on a farmerless farm.
Made from a series of tall, rectangular metal
frames secured with wooden cross-supports
and flooring, it turned in place like a merry-
go-round on wheels built with spare parts

from a Model T Ford. Jansky had tuned the hundred-foot-long contraption to a wavelength of about fifteen meters, corresponding to a frequency of 20.5 megahertz.[†] Jansky's agenda, on behalf of his employer, Bell Telephone Laboratories, was to study any hisses from Earth-based radio sources that might contaminate terrestrial radio communications. This greatly resembles the task that Bell Labs gave Penzias and Wilson, thirty-five years later, to find microwave noise in their receiver, as we saw in chapter 3, which led to the discovery of the cosmic microwave background.

By spending a couple of years painstakingly tracking and timing the static hiss that registered on his jury-rigged antenna, Jansky had discovered that radio waves emanate not just from local thunderstorms and other known terrestrial sources, but also from the center of

---

[†] All waves follow the simple equation: *speed = frequency × wavelength*. At a constant speed, if you increase the wavelength, the wave itself will have smaller frequency, and vice versa, so that when you multiply the two quantities you recover the same speed of the wave every time. Works for light, sound, and even fans doing the "Wave" at sports arenas—anything that's a traveling wave.

the Milky Way galaxy. That region of the sky swung by the telescope's field of view every twenty-three hours and fifty-six minutes: exactly the period of Earth's rotation in space and thus exactly the time needed to return the galactic center to the same angle and elevation on the sky. Karl Jansky published his results under the title "Electrical Disturbances Apparently of Extraterrestrial Origin."[†]

With that observation, radio astronomy was born—but minus Jansky himself. Bell Labs retasked him, preventing him from pursuing the fruits of his own seminal discovery. A few years later, though, a self-starting American named Grote Reber, from Wheaton, Illinois, built a thirty-foot-wide, metal-dish radio telescope in his own backyard. In 1938, under nobody's employ, Reber confirmed Jansky's discovery, and spent the next five years making low-resolution maps of the radio sky.

Reber's telescope was small and crude by today's standards. Modern radio telescopes are

[†] Karl Jansky, "Electrical Disturbances Apparently of Extraterrestrial Origin," *Proceedings of the Institute for Radio Engineers* 21, no. 10 (1933): 1387.

quite another matter. Unbound by backyards, they're sometimes downright humongous. MK 1, which began its working life in 1957, is the planet's first genuinely gigantic radio telescope—a single, steerable, 250-foot-wide, solid-steel dish at the Jodrell Bank Observatory near Manchester, England. A couple of months after MK 1 opened for business, the Soviet Union launched *Sputnik 1*, and Jodrell Bank's dish suddenly became just the thing to track the little orbiting hunk of hardware—making it the forerunner of today's Deep Space Network for tracking planetary space probes.

The world's largest radio telescope, completed in 2016, is called the Five-hundred-meter Aperture Spherical radio Telescope, or "FAST" for short. It was built by China in their Guizhou Province, and is larger in area than thirty football fields. If aliens ever give us a call, the Chinese will be the first to know.

*

Another variety of radio telescope is the interferometer, comprising arrays of identical dish antennas, spread across swaths of countryside and electronically linked to work in concert. The result is a single, coherent, super-high-resolution image of radio-emitting cosmic objects. Although "supersize me" was the unwritten motto for telescopes long before the fast food industry coined the slogan, radio interferometers form a jumbo class unto themselves. One of them, a very large array of radio dishes near Socorro, New Mexico, is officially called the Very Large Array, with twenty-seven eighty-two-foot dishes positioned on tracks crossing twenty-two miles of desert plains. This observatory is so cosmogenic, it has appeared as a backdrop in the films *2010: The Year We Make Contact* (1984), *Contact* (1997), and *Transformers* (2007). There's also the Very Long Baseline Array, with ten eighty-two-foot dishes spanning 5,000 miles from Hawaii to the Virgin Islands, enabling the highest resolution of any radio telescope in the world.

In the microwave band, relatively new to interferometers, we've got the sixty-six antennas of ALMA, the Atacama Large Millimeter Array, in the remote Andes Mountains of northern Chile. Tuned for wavelengths that range from fractions of a millimeter to several centimeters, ALMA gives astrophysicists high-resolution access to categories of cosmic action unseen in other bands, such as the structure of collapsing gas clouds as they become nurseries from which stars are born. ALMA's location is, by intention, the most arid landscape on Earth—three miles above sea level and well above the wettest clouds. Water may be fine for microwave cooking but it's bad for astrophysicists, because the water vapor in Earth's atmosphere chews up pristine microwave signals from across the galaxy and beyond. These two phenomena are, of course, related: water is the most common ingredient in food, and microwave ovens primarily heat water. Taken together, you get the best indication that water absorbs microwave frequencies. So if you want clean observations of cosmic objects, you must minimize the amount of water vapor between your telescope and the universe, just as ALMA has done.

✳

At the ultrashort-wavelength end of the electromagnetic spectrum you find the high-frequency, high-energy gamma rays, with wavelengths measured in picometers.[†] Discovered in 1900, they were not detected from space until a new kind of telescope was placed aboard NASA's *Explorer XI* satellite in 1961.

Anybody who watches too many sci-fi movies knows that gamma rays are bad for you. You might turn green and muscular, or spiderwebs might squirt from your wrists. But they're also hard to trap. They pass right through ordinary lenses and mirrors. How, then, to observe them? The guts of *Explorer XI*'s telescope held a device called a scintillator, which responds to incoming gamma rays by pumping out electrically charged particles. If you measure energies of the particles, you can tell what kind of high-energy light created them.

Two years later the Soviet Union, the United Kingdom, and the United States signed the Limited Test Ban Treaty, which prohibited nuclear testing underwater, in the atmosphere, and in space—where nuclear fallout could spread and contaminate places outside your own country's perimeter. But this was the Cold

[†] "pico-" is the metric prefix for one-trillionth.

War, a time when nobody believed anybody about anything. Invoking the military edict "trust but verify," the U.S. deployed a new series of satellites, the *Velas*, to scan for gamma ray bursts that would result from Soviet nuclear tests. The satellites indeed found bursts of gamma rays, almost daily. But Russia wasn't to blame. These came from deep space—and were later shown to be the calling card of intermittent, distant, titanic stellar explosions across the universe, signaling the birth of gamma ray astrophysics, a new branch of study in my field.

In 1994, NASA's Compton Gamma Ray Observatory detected something as unexpected as the *Velas*' discoveries: frequent flashes of gamma rays right near Earth's surface. They were sensibly dubbed "terrestrial gamma-ray flashes." Nuclear holocaust? No, as is evident from the fact that you're reading this sentence. Not all bursts of gamma rays are equally lethal, nor are they all of cosmic origin. In this case, at least fifty flashes emanate daily near the tops of thunderclouds, a split second before ordinary lightning bolts strike. Their origin remains a bit of a mystery, but the best explanation holds that in the electrical storm, free electrons accelerate to near the speed of light and then slam into the nuclei of atmospheric atoms, generating gamma rays.

*

Today, telescopes operate in every invisible part of the spectrum, some from the ground but most from space, where a telescope's view is unimpeded by Earth's absorptive atmosphere. We can now observe phenomena ranging from low-frequency radio waves a dozen meters long, crest to crest, to high-frequency gamma rays no longer than a quadrillionth of a meter. That rich palette of light supplies no end of astrophysical discoveries: Curious how much gas lurks among the stars in galaxies? Radio telescopes do that best. There is no knowledge of the cosmic background, and no real understanding of the big bang, without microwave telescopes. Want to peek at stellar nurseries deep inside galactic gas clouds? Pay attention to what infrared telescopes do. How about emissions from the vicinity of ordinary black holes and supermassive black holes in the center of a galaxy? Ultraviolet and X-ray telescopes do that best. Want to watch the high-energy explosion of a giant star, whose mass is as great as forty suns? Catch the drama via gamma ray telescopes.

We've come a long way since Herschel's experiments with rays that were "unfit for vision," empowering us to explore the universe for what it is, rather than for what it seems to be. Herschel would be proud. We achieved true cosmic vision only after seeing the unsee-able: a dazzlingly rich collection of objects and phenomena across space and across time that we may now dream of in our philosophy.

10.

Between the Planets

From a distance, our solar system looks empty. If you enclosed it within a sphere—one large enough to contain the orbit of Neptune, the outermost planet[†]—then the volume occupied by the Sun, all planets, and their moons would take up a little more than one-trillionth the enclosed space. But it's not empty, the space between the planets contains all manner of chunky rocks, pebbles, ice balls, dust, streams of charged particles, and far-flung probes. The space is also permeated by monstrous gravitational and magnetic fields.

Interplanetary space is so not-empty that

[†] No, it's not Pluto. Get over it.

Earth, during its 30 kilometer-per-second orbital journey, plows through hundreds of tons of meteors per day—most of them no larger than a grain of sand. Nearly all of them burn in Earth's upper atmosphere, slamming into the air with so much energy that the debris vaporizes on contact. Our frail species evolved under this protective blanket. Larger, golf-ball-size meteors heat fast but unevenly, and often shatter into many smaller pieces before they vaporize. Still larger meteors singe their surface but otherwise make it all the way to the ground intact. You'd think that by now, after 4.6 billion trips around the Sun, Earth would have "vacuumed" up all possible debris in its orbital path. But things were once much worse. For a half-billion years after the formation of the Sun and its planets, so much junk rained down on Earth that heat from the persistent energy of impacts rendered Earth's atmosphere hot and our crust molten.

One substantial hunk of junk led to the formation of the Moon. The unexpected scarcity of iron and other higher-mass elements in the Moon, derived from lunar samples returned by Apollo astronauts, indicates that the Moon most likely burst forth from Earth's iron-poor crust and mantle after a glancing collision with a wayward Mars-sized proto-planet. The orbiting debris from this encounter coalesced to form our lovely, low-density satellite. Apart from this newsworthy event, the period of heavy bombardment that Earth endured during its infancy was not unique among the planets and other large bodies of the solar system. They each sustained similar damage, with the airless, erosionless surfaces of the Moon and Mercury preserving much of the cratered record from this period.

Not only is the solar system scarred by the flotsam of its formation, but nearby interplanetary space also contains rocks of all sizes that were jettisoned from Mars, the Moon, and Earth by the ground's recoil from high-speed impacts. Computer studies of meteor strikes demonstrate conclusively that surface rocks near impact zones can get thrust upward with enough speed to escape the body's gravitational tether. At the rate we are discovering meteorites on Earth whose origin is Mars, we conclude that about a thousand tons of Martian rocks rain down on Earth each year. Perhaps the same amount reaches Earth from the Moon. In retrospect, we didn't have to go to the Moon to retrieve Moon rocks. Plenty come to us, although they were not of our choosing and we didn't yet know it during the Apollo program.

*

Most of the solar system's asteroids live and work in the main asteroid belt, a roughly flat zone between the orbits of Mars and Jupiter. By tradition, the discoverers get to name their asteroids whatever they like. Often drawn by artists as a region of cluttered, meandering rocks in the plane of the solar system, the asteroid belt's total mass is less than five percent that of the Moon, which is itself barely more than one percent of Earth's mass. Sounds insignificant. But accumulated perturbations of their orbits continually create a deadly subset, perhaps a few thousand, whose eccentric paths intersect Earth's orbit. A simple calculation reveals that most of them will hit Earth within a hundred million years. The ones

larger than about a kilometer across will collide with enough energy to destabilize Earth's
ecosystem and put most of Earth's land species
at risk of extinction.

That would be bad.

Asteroids are not the only space objects that
pose a risk to life on Earth. The Kuiper belt is
a comet-strewn swath of circular real estate
that begins just beyond the orbit of Neptune,
includes Pluto, and extends perhaps as far
again from Neptune as Neptune is from the
Sun. The Dutch-born American astronomer
Gerard Kuiper advanced the idea that in the
cold depths of space, beyond the orbit of Neptune, there reside frozen leftovers from the formation of the solar system. Without a massive
planet upon which to fall, most of these comets will orbit the Sun for billions more years.
As is true for the asteroid belt, some objects of
the Kuiper belt travel on eccentric paths that
cross the orbits of other planets. Pluto and
its ensemble of siblings called Plutinos cross
Neptune's path around the Sun. Other Kuiper
belt objects plunge all the way down to the
inner solar system, crossing planetary orbits

with abandon. This subset includes Halley, the most famous comet of them all.

Far beyond the Kuiper belt, extending half-way to the nearest stars, lives a spherical reservoir of comets called the Oort cloud, named for Jan Oort, the Dutch astrophysicist who first deduced its existence. This zone is responsible for the long-period comets, those with orbital periods far longer than a human lifetime. Unlike Kuiper belt comets, Oort cloud comets can rain down on the inner solar system from any angle and from any direction. The two brightest of the 1990s, comets Hale-Bopp and Hyakutake, were both from the Oort cloud and are not coming back anytime soon.

*

If we had eyes that could see magnetic fields, Jupiter would look ten times larger than the full Moon in the sky. Spacecraft that visit Jupiter must be designed to remain unaffected by this powerful force. As the English physicist Michael Faraday demonstrated in the 1800s, if you pass a wire across a magnetic field you generate a voltage difference along the wire's length. For this reason, fast-moving metal space probes will have electrical currents induced within them. Meanwhile, these currents generate magnetic fields of their own that interact with the ambient magnetic field in ways that retard the space probe's motion.

Last I had kept count, there were fifty-six moons among the planets in the solar system. Then I woke up one morning to learn that another dozen had been discovered around Saturn. After that incident, I decided to no longer keep track. All I care about now is whether any of them would be fun places to visit or to study. By some measures, the solar system's moons are much more fascinating than the planets they orbit.

*

Earth's Moon is about 1/400th the diameter of the Sun, but it is also 1/400th as far from us, making the Sun and the Moon the same size on the sky—a coincidence not shared by any other planet–moon combination in the solar system, allowing for uniquely photogenic total solar eclipses. Earth has also tidally locked the Moon, leaving it with identical periods of rotation on its axis and revolution around Earth. Wherever and whenever this happens, the locked moon shows only one face to its host planet.

Jupiter's system of moons is replete with oddballs. Io, Jupiter's closest moon, is tidally locked and structurally stressed by interactions with Jupiter and with other moons, pumping enough heat into the little orb to render molten its interior rocks; Io is the most volcanically active place in the solar system. Jupiter's moon Europa has enough H_2O that its heating mechanism—the same one at work on Io—has melted the subsurface ice, leaving a warmed ocean below. If ever there was a next-best place to look for life, it's here. (An artist coworker of mine once asked whether alien life forms from Europa would be called Europeans. The absence of any other plausible answer forced me to say yes.)

Pluto's largest moon, Charon, is so big and close to Pluto that Pluto and Charon have each tidally locked the other: their rotation periods and their periods of revolution are identical. We call this a "double tidal lock," which sounds like a yet-to-be-invented wrestling hold.

By convention, moons are named for Greek personalities in the life of the Greek counterpart to the Roman god after whom the planet itself was named. The classical gods led complicated social lives, so there is no shortage of characters from which to draw. The lone exception to this rule applies to the moons of Uranus, which are named for assorted protagonists in British lit. William Herschel was the first person to discover a planet beyond those easily visible to the naked eye, and he was ready to name it after the King, under whom he faithfully served. Had Herschel succeeded, the planet list would read: Mercury, Venus, Earth, Mars, Jupiter, Saturn, and George. Fortunately, clearer heads prevailed and the classical name Uranus was adopted some years later. But his original suggestion to name the moons after characters in William Shakespeare's plays and Alexander Pope's poems remains the tradition to this day. Among its twenty-seven moons we

find Ariel, Cordelia, Desdemona, Juliet, Oph-
elia, Portia, Puck, Umbriel, and Miranda.

The Sun loses material from its surface at a
rate of more than a million tons per second. We
call this the "solar wind," which takes the form
of high-energy charged particles. Traveling up
to a thousand miles per second, these particles
stream through space and are deflected by
planetary magnetic fields. The particles spiral down toward the north and south magnetic
poles, forcing collisions with gas molecules and
leaving the atmosphere aglow with colorful
aurora. The Hubble Space Telescope has spot-
ted aurora near the poles of both Saturn and
Jupiter. And on Earth, the aurora borealis and
australis (the northern and southern lights)
serve as intermittent reminders of how nice it
is to have a protective atmosphere.

Earth's atmosphere is commonly described as
extending dozens of miles above Earth's surface.
Satellites in "low" Earth orbit typically travel
between one hundred and four hundred miles
up, completing an orbit in about ninety minutes. While you can't breathe at those altitudes,
some atmospheric molecules remain—enough

to slowly drain orbital energy from unsuspect-
ing satellites. To combat this drag, satellites in
low orbit require intermittent boosts, lest they
fall back to Earth and burn up in the atmosphere. An alternative way to define the edge of
our atmosphere is to ask where its density of gas
molecules equals the density of gas molecules
in interplanetary space. Under that definition,
Earth's atmosphere extends thousands of miles.

Orbiting high above this level, twenty-three
thousand miles up (one-tenth of the distance to
the Moon) are the communications satellites.
At this special altitude, Earth's atmosphere is
not only irrelevant, but the speed of the satellite is low enough for it to require a full day to
complete one revolution around Earth. With
an orbit precisely matching the rotation rate of
Earth, these satellites appear to hover, which
make them ideal for relaying signals from one
part of Earth's surface to another.

*

Newton's laws specifically state that, while
the gravity of a planet gets weaker and weaker
the farther from it you travel, there is no dis-

tance where the force of gravity reaches zero. The planet Jupiter, with its mighty gravitational field, bats out of harm's way many comets that would otherwise wreak havoc on the inner solar system. Jupiter acts as a gravitational shield for Earth, a burly big brother, allowing long (hundred-million-year) stretches of relative peace and quiet on Earth. Without Jupiter's protection, complex life would have a hard time becoming interestingly complex, always living at risk of extinction from a devastating impact.

We have exploited the gravitational fields of planets for nearly every probe launched into space. The *Cassini* probe, for example, which visited Saturn, was gravitationally assisted twice by Venus, once by Earth (on a return flyby), and once by Jupiter. Like a multi-cushion billiard shot, trajectories from one planet to another are common. Our tiny probes would not otherwise have enough speed and energy from our rockets to reach their destination.

I am now accountable for some of the solar system's interplanetary debris. In November 2000, the main-belt asteroid 1994KA, discovered by David Levy and Carolyn Shoemaker, was named 13123–Tyson in my honor. While I enjoyed the distinction, there's no particular reason to get big-headed about it; plenty of asteroids have familiar names such as Jody, Harriet, and Thomas. There are even asteroids out there named Merlin, James Bond, and Santa. Now in the hundreds of thousands, the asteroid count might soon challenge our capacity to name them. Whether or not that day arrives, I take comfort knowing that my chunk of cosmic debris is not alone as it litters the space between the planets, being joined by a long list of other chunks named for real and fictional people.

I'm also glad that, at the moment, my asteroid is not headed towards Earth.

11.

Exoplanet Earth

Whether you prefer to sprint, swim, walk, or crawl from one place to another on Earth, you can enjoy close-up views of our planet's unlimited supply of things to notice. You might see a vein of pink limestone on the wall of a canyon, a ladybug eating an aphid on the stem of a rose, a clamshell poking out from the sand. All you have to do is look.

From the window of an ascending jetliner, those surface details rapidly disappear. No aphid appetizers. No curious clams. Reach cruising altitude, around seven miles up, and identifying major roadways becomes a challenge.

Detail continues to vanish as you rise into space. From the window of the International

Space Station, which orbits at about 250 miles up, you might find Paris, London, New York, and Los Angeles in the daytime, but only because you learned where they are in geography class. At night, their sprawling cityscapes present an obvious glow. By day, contrary to common wisdom, you probably won't see the Great Pyramids at Giza, and you certainly won't see the Great Wall of China. Their obscurity is partly the result of having been made from the soil and stone of the surrounding landscape. And although the Great Wall is thousands of miles long, it's only about twenty feet wide—much narrower than the U.S. interstate highways you can barely see from a transcontinental jet.

From orbit, with the unaided eye, you would have seen smoke plumes rising from the oilfield fires in Kuwait at the end of the first Persian Gulf War in 1991 and smoke from the burning World Trade Center towers in New York City on September 11, 2001. You will also notice the green–brown boundaries between swaths of irrigated and arid land. Beyond that shortlist, there's not much else made by humans

that's identifiable from hundreds of miles up
in the sky. You can see plenty of natural scen-
ery, though, including hurricanes in the Gulf
of Mexico, ice floes in the North Atlantic, and
volcanic eruptions wherever they occur.

From the Moon, a quarter million miles
away, New York, Paris, and the rest of Earth's
urban glitter doesn't even show up as a twin-
kle. But from your lunar vantage you can still
watch major weather fronts move across the
planet. From Mars at its closest, some thirty-
five million miles away, massive snow-capped
mountain chains and the edges of Earth's con-
tinents would be visible through a large backyard telescope. Travel out to Neptune, three
billion miles away—just down the block on
a cosmic scale—and the Sun itself becomes
a thousand times dimmer, now occupying a
thousandth the area on the daytime sky that
it occupies when seen from Earth. And what
of Earth itself? It's a speck no brighter than a
dim star, all but lost in the glare of the Sun.

A celebrated photograph taken in 1990
from just beyond Neptune's orbit by the *Voyager 1* spacecraft shows just how underwhelm-

ing Earth looks from deep space: a "pale blue dot," as the American astrophysicist Carl Sagan called it. And that's generous. Without the help of a caption, you might not even know it's there.

What would happen if some big-brained aliens from the great beyond scanned the skies with their naturally superb visual organs, further aided by alien state-of-the-art optical accessories? What visible features of planet Earth might they detect?

Blueness would be first and foremost. Water covers more than two thirds of Earth's surface; the Pacific Ocean alone spans nearly an entire side of the planet. Any beings with enough equipment and expertise to detect our planet's color would surely infer the presence of water, the third most abundant molecule in the universe.

If the resolution of their equipment were high enough, the aliens would see more than just a pale blue dot. They would see intricate coastlines, too, strongly suggesting that the water is liquid. And smart aliens would surely know that if a planet has liquid water, then the planet's temperature and atmospheric pressure fall within a well-determined range.

Earth's distinctive polar ice caps, which grow and shrink from the seasonal temperature variations, could also be seen using visible light. So could our planet's twenty-four-hour rotation, because recognizable landmasses rotate into view at predictable intervals of time. The aliens would also see major weather systems come and go; with careful study, they could readily distinguish features related to clouds in the atmosphere from features related to the surface of Earth itself.

Time for a reality check. The nearest exoplanet—the nearest planet in orbit around a star that is not the Sun—can be found in our neighbor star system Alpha Centauri, about four light-years from us and visible mostly from our southern hemisphere. Most of the cataloged exoplanets lie from dozens up to hundreds of light-years away. Earth's brightness is less than one-billionth that of the Sun, and our planet's proximity to the Sun would make it extremely hard for anybody to see Earth directly with a visible light telescope. It's like trying to detect the light of a firefly in the vicinity of a Hollywood searchlight. So if aliens have found us, they are likely looking in wavelengths other than visible light, like infrared, where our brightness relative to the Sun is a bit better than in visible light—or else their engineers are adapting some other strategy altogether.

Maybe they're doing what some of our own planet-hunters typically do: monitoring stars to see if they jiggle at regular intervals. A star's periodic jiggle betrays the existence of an orbiting planet that may be too dim to see directly. Contrary to what most people suppose, a planet does not orbit its host star. Instead, both the planet and its host star revolve around their common center of mass. The more massive the planet, the larger the star's response must be, and the more measurable the jiggle gets when you analyze the star's light. Unfortunately for planet-hunting aliens, Earth is puny, so the Sun barely budges, which would further challenge alien engineers.

*

NASA's Kepler telescope, designed and tuned to discover Earth-like planets around Sun-like stars, invoked yet another method of detection, mightily adding to the exoplanet catalog. Kepler searched for stars whose total brightness drops slightly, and at regular intervals. In these cases, Kepler's line of sight is just right to see a star get dimmer, by a tiny fraction, due to one of its own planets crossing directly in front of the host star. With this method, you can't see the planet itself. You can't even see any features on the star's surface. Kepler simply tracked changes in a star's total light, but added thousands of exoplanets to the catalog, including hundreds of multiplanet star systems. From these data, you also learn the size of the exoplanet, its orbital period, and its orbital distance from the host star. You can also make an educated inference on the planet's mass.

If you're wondering, when Earth passes in front of the Sun—which is always happening for some line of sight in the galaxy—we block 1/10,000th of the Sun's surface, thereby briefly dimming the Sun's total light by 1/10,000th of its normal brightness. Fine as it goes. They'll discover that Earth exists, but learn nothing about happenings on Earth's surface.

Radio waves and microwaves might work. Maybe our eavesdropping aliens have something like the 500-meter radio telescope in the Guizhou province of China. If they do, and if they tune to the right frequencies, they'll certainly notice Earth—or rather, they'll notice our modern civilization as one of the most luminous sources in the sky. Consider everything we've got that generates radio waves and microwaves: not only traditional radio itself, but also broadcast television, mobile phones, microwave ovens, garage-door openers, car-door unlockers, commercial radar, military radar, and communications satellites. We're ablaze in long-frequency waves—spectacular evidence that something unusual is going on here, because in their natural state, small rocky planets emit hardly any radio waves at all.

So if those alien eavesdroppers turn their own version of a radio telescope in our direction, they might infer that our planet hosts technology. One complication, though: other interpretations are possible. Maybe they wouldn't be able to distinguish Earth's signals from those of the larger planets in our solar system, all of which are sizable sources of radio waves, especially Jupiter. Maybe they'd think

we were a new kind of odd, radio-intensive
planet. Maybe they wouldn't be able to distinguish Earth's radio emissions from those of
the Sun, forcing them to conclude that the Sun
is a new kind of odd, radio-intensive star.

Astrophysicists right here on Earth, at the
University of Cambridge in England, were
similarly stumped back in 1967. While sur-
veying the skies with a radio telescope for any
source of strong radio waves, Antony Hewish
and his team discovered something extremely
odd: an object pulsing at precise, repeating
intervals of slightly more than a second. Jocelyn Bell, a graduate student of Hewish's at
the time, was the first to notice it.

Soon Bell's colleagues established that the
pulses came from a great distance. The thought
that the signal was technological—another
culture beaming evidence of its activities
across space—was irresistible. As Bell recounts,
"We had no proof that it was an entirely natural radio emission. . . . Here was I trying to
get a Ph.D. out of a new technique, and some
silly lot of little green men had to choose my
aerial and my frequency to communicate with

us."† Within a few days, however, she discovered other repeating signals coming from other places in our Milky Way galaxy. Bell and her associates realized they'd discovered a new class of cosmic object—a star made entirely of neutrons that pulses with radio waves for every rotation it executes. Hewish and Bell sensibly named them "pulsars."

Turns out, intercepting radio waves isn't the only way to be snoopy. There's also cosmochemistry. The chemical analysis of planetary atmospheres has become a lively field of modern astrophysics. As you might guess, cosmochemistry depends on spectroscopy—the analysis of light by means of a spectrometer. By exploiting the tools and tactics of spectroscopists, cosmochemists can infer the presence of life on an exoplanet, regardless of whether that life has sentience, intelligence, or technology.

The method works because every element, every molecule—no matter where it exists in the universe—absorbs, emits, reflects, and

---

† Jocelyn Bell, *Annals of the New York Academy of Sciences* 302 (1977): 685.

scatters light in a unique way. And as already discussed, pass that light through a spectrometer, and you'll find features that can rightly be called chemical fingerprints. The most visible fingerprints are made by the chemicals most excited by the pressure and temperature of their environment. Planetary atmospheres are rich with such features. And if a planet is teeming with flora and fauna, its atmosphere will be rich with biomarkers—spectral evidence of life. Whether biogenic (produced by any or all life-forms), anthropogenic (produced by the widespread species *Homo sapiens*), or technogenic (produced only by technology), such rampant evidence will be hard to conceal.

Unless they happen to be born with built-in spectroscopic sensors, our space-snooping aliens would need to build a spectrometer to read our fingerprints. But above all, Earth would have to cross in front of the Sun (or some other source), permitting light to pass through our atmosphere and continue on to the aliens. That way, the chemicals in Earth's atmosphere could interact with the light, leaving their marks for all to see.

Some molecules—ammonia, carbon dioxide, water—show up abundantly in the universe, whether life is present or not. But other molecules thrive in the presence of life itself. Another readily detected biomarker is Earth's sustained level of the molecule methane, two-thirds of which is produced by human-related activities such as fuel oil production, rice cultivation, sewage, and the burps and farts of domestic livestock. Natural sources, comprising the remaining third, include decomposing vegetation in wetlands and termite effluences. Meanwhile, in places where free oxygen is scarce, methane does not always require life to form. At this very moment, astrobiologists are arguing over the exact origin of trace methane on Mars and the copious quantities of methane on Saturn's moon Titan, where cows and termites we presume do not dwell.

If the aliens track our nighttime side while we orbit our host star, they might notice a surge of sodium from the widespread use of sodium-vapor streetlights that switch on at dusk in urban and suburban municipalities. Most telling, however, would be all our free-floating oxygen, which constitutes a full fifth of our atmosphere.

Oxygen—which, after hydrogen and helium, is the third most abundant element in the cosmos—is chemically active and bonds readily with atoms of hydrogen, carbon, nitrogen, silicon, sulfur, iron, and so on. It even bonds with itself. Thus, for oxygen to exist in a steady state, something must be liberating it as fast as it's being consumed. Here on Earth, the liberation is traceable to life. Photosynthesis, carried out by plants and many bacteria, creates free oxygen in the oceans and in the atmosphere. Free oxygen, in turn, enables the existence of oxygen-metabolizing life, including us and practically every other creature in the animal kingdom.

We Earthlings already know the significance of our planet's distinctive chemical fingerprints. But distant aliens who come upon us will have to interpret their findings and test their assumptions. Must the periodic appearance of sodium be technogenic? Free oxygen is surely biogenic. How about methane? It, too, is chemically unstable, and yes, some of it is anthropogenic, but as we've seen, methane has nonliving agents as well.

If the aliens decide that Earth's chemical features are sure evidence of life, maybe they'll wonder if the life is intelligent. Presumably the aliens communicate with one another, and perhaps they'll presume that other intelligent life-forms communicate, too. Maybe that's when they'll decide to eavesdrop on Earth with their radio telescopes to see what part of the electromagnetic spectrum its inhabitants have mastered. So, whether the aliens explore with chemistry or with radio waves, they might come to the same conclusion: a planet where there's advanced technology must be populated with intelligent life-forms, who may occupy themselves discovering how the universe works and how to apply its laws for personal or public gain.

Looking more closely at Earth's atmospheric fingerprints, human biomarkers will also include sulfuric, carbonic, and nitric acids, and other components of smog from the burning of fossil fuels. If the curious aliens happen to be socially, culturally, and technologically

more advanced than we are, then they will
surely interpret these biomarkers as convincing evidence for the absence of intelligent life
on Earth.

*

The first exoplanet was discovered in 1995,
and, as of this writing, the tally is rising
through three thousand, most found in a small
pocket of the Milky Way around the solar system. So there's plenty more where they came
from. After all, our galaxy contains more than
a hundred billion stars, and the known universe harbors some hundred billion galaxies.

Our search for life in the universe drives the
search for exoplanets, some of which resemble
Earth—not in detail, of course, but in overall properties. Latest estimates, extrapolating
from the current catalogs, suggest as many as
forty billion Earth-like planets in the Milky
Way alone. Those are the planets our descendants might want to visit someday, by choice,
if not by necessity.

12.

Reflections on the Cosmic Perspective

Of all the sciences cultivated by mankind, Astronomy is acknowledged to be, and undoubtedly is, the most sublime, the most interesting, and the most useful. For, by knowledge derived from this science, not only the bulk of the Earth is discovered ...; but our very faculties are enlarged with the grandeur of the ideas it conveys, our minds exalted above [their] low contracted prejudices.

JAMES FERGUSON, 1757[†]

[†] James Ferguson, *Astronomy Explained Upon Sir Isaac Newton's Principles, And Made Easy To Those Who Have Not Studied Mathematics* (London, 1757).

Long before anyone knew that the universe had a beginning, before we knew that the nearest large galaxy lies two million light-years from Earth, before we knew how stars work or whether atoms exist, James Ferguson's enthusiastic introduction to his favorite science rang true. Yet his words, apart from their eighteenth-century flourish, could have been written yesterday.

But who gets to think that way? Who gets to celebrate this cosmic view of life? Not the migrant farmworker. Not the sweatshop worker. Certainly not the homeless person rummaging through the trash for food. You need the luxury of time not spent on mere survival. You need to live in a nation whose government values the search to understand humanity's place in the universe. You need a society in which intellectual pursuit can take you to the frontiers of discovery, and in which news of your discoveries can be routinely disseminated. By those measures, most citizens of industrialized nations do quite well.

Yet the cosmic view comes with a hidden cost. When I travel thousands of miles to spend a few moments in the fast-moving shadow of the Moon during a total solar eclipse, sometimes I lose sight of Earth.

When I pause and reflect on our expanding universe, with its galaxies hurtling away from one another, embedded within the ever-stretching, four-dimensional fabric of space and time, sometimes I forget that uncounted people walk this Earth without food or shelter, and that children are disproportionately represented among them.

When I pore over the data that establish the mysterious presence of dark matter and dark energy throughout the universe, sometimes I forget that every day—every twenty-four-hour rotation of Earth—people kill and get killed in the name of someone else's conception of God, and that some people who do not kill in the name of God, kill in the name of needs or wants of political dogma.

When I track the orbits of asteroids, comets, and planets, each one a pirouetting dancer in a cosmic ballet, choreographed by the forces of gravity, sometimes I forget that too many people act in wanton disregard for the delicate interplay of Earth's atmosphere, oceans, and land, with consequences that our children and our children's children will witness and pay for with their health and well-being.

And sometimes I forget that powerful people rarely do all they can to help those who cannot help themselves.

I occasionally forget those things because, however big the world is—in our hearts, our minds, and our outsized digital maps—the universe is even bigger. A depressing thought to some, but a liberating thought to me.

Consider an adult who tends to the traumas of a child: spilled milk, a broken toy, a scraped knee. As adults we know that kids have no clue of what constitutes a genuine problem, because inexperience greatly limits their childhood perspective. Children do not yet know that the world doesn't revolve around them.

As grown-ups, dare we admit to ourselves that we, too, have a collective immaturity of view? Dare we admit that our thoughts and behaviors spring from a belief that the world revolves around us? Apparently not. Yet evidence abounds. Part the curtains of society's racial, ethnic, religious, national, and cultural conflicts, and you find the human ego turning the knobs and pulling the levers.

Now imagine a world in which everyone, but especially people with power and influence, holds an expanded view of our place in the cosmos. With that perspective, our problems would shrink—or never arise at all--and we could celebrate our earthly differences while shunning the behavior of our predecessors who slaughtered one another because of them.

*

Back in January 2000, the newly rebuilt Hayden Planetarium in New York City featured a space show titled *Passport to the Universe*,[†] which took visitors on a virtual zoom from the planetarium out to the edge of the cosmos. En route, the audience viewed Earth, then the

[†] *Passport to the Universe* was written by Ann Druyan and Steven Soter, who are also the co-authors of the 2014 Fox miniseries *Cosmos: A SpaceTime Odyssey*, hosted by this author. They also teamed up with Carl Sagan on the original 1980 PBS miniseries *Cosmos: A Personal Voyage*.

solar system, then watched the hundred billion stars of the Milky Way galaxy shrink, in
turn, to barely visible dots on the planetarium's dome.

Within a month of opening day, I received
a letter from an Ivy League professor of psy-
chology whose expertise was in things that
make people feel insignificant. I never knew
one could specialize in such a field. He wanted
to administer a before-and-after questionnaire
to visitors, assessing the depth of their depres-
sion after viewing the show. *Passport to the
Universe*, he wrote, elicited the most dramatic
feelings of smallness and insignificance he
had ever experienced.

How could that be? Every time I see the
space show (and others we've produced), I feel
alive and spirited and connected. I also feel
large, knowing that the goings-on within the
three-pound human brain are what enabled us
to figure out our place in the universe.

Allow me to suggest that it's the professor,
not I, who has misread nature. His ego was
unjustifiably big to begin with, inflated by
delusions of significance and fed by cultural

assumptions that human beings are more important than everything else in the universe.

In all fairness to the fellow, powerful forces in society leave most of us susceptible. As was I, until the day I learned in biology class that more bacteria live and work in one centimeter of my colon than the number of people who have ever existed in the world. That kind of information makes you think twice about who—or what—is actually in charge.

From that day on, I began to think of people not as the masters of space and time but as participants in a great cosmic chain of being, with a direct genetic link across species both living and extinct, extending back nearly four billion years to the earliest single-celled organisms on Earth.

I know what you're thinking: we're smarter than bacteria.

No doubt about it, we're smarter than every other living creature that ever ran, crawled, or slithered on Earth. But how smart is that? We cook our food. We compose poetry and music. We do art and science. We're good at math. Even if you're bad at math, you're probably much better at it than the smartest chimpanzee, whose genetic identity varies in only trifling ways from ours. Try as they might, primatologists will never get a chimpanzee to do long division, or trigonometry.

If small genetic differences between us and our fellow apes account for what appears to be a vast difference in intelligence, then maybe that difference in intelligence is not so vast after all.

Imagine a life-form whose brainpower is to ours as ours is to a chimpanzee's. To such a species, our highest mental achievements would be trivial. Their toddlers, instead of learning their ABCs on Sesame Street, would learn multivariable calculus on Boolean Boulevard.[†] Our most complex theorems, our deepest philosophies, the cherished works of our most creative artists, would be projects their schoolkids bring home for Mom and Dad to display on

---

[†] Boolean algebra is a branch of mathematics that addresses values of true or false in its variables, commonly represented by 0 and 1, and is foundational to the world of computing. Named for the eighteenth-century English mathematician George Boole.

the refrigerator door with a magnet. These creatures would study Stephen Hawking (who occupies the same endowed professorship once held by Isaac Newton at the University of Cambridge) because he's slightly more clever than other humans. Why? He can do theoretical astrophysics and other rudimentary calculations in his head, like their little Timmy who just came home from alien preschool.

If a huge genetic gap separated us from our closest relative in the animal kingdom, we could justifiably celebrate our brilliance. We might be entitled to walk around thinking we're distant and distinct from our fellow creatures. But no such gap exists. Instead, we are one with the rest of nature, fitting neither above nor below, but within.

Need more ego softeners? Simple comparisons of quantity, size, and scale do the job well.

Take water. It's common, and vital. There are more molecules of water in an eight-ounce cup of the stuff than there are cups of water in all the world's oceans. Every cup that passes through a single person and eventually rejoins the world's water supply holds enough molecules to mix 1,500 of them into every other cup of water in the world. No way around it: some of the water you just drank passed through the kidneys of Socrates, Genghis Khan, and Joan of Arc.

How about air? Also vital. A single breathful draws in more air molecules than there are breathfuls of air in Earth's entire atmosphere. That means some of the air you just breathed passed through the lungs of Napoleon, Beethoven, Lincoln, and Billy the Kid.

Time to get cosmic. There are more stars in the universe than grains of sand on any beach, more stars than seconds have passed since Earth formed, more stars than words and sounds ever uttered by all the humans who ever lived.

Want a sweeping view of the past? Our unfolding cosmic perspective takes you there. Light takes time to reach Earth's observatories from the depths of space, and so you see objects and phenomena not as they are but as they once were, back almost to the beginning of time itself. Within that horizon of reckoning, cosmic evolution unfolds continuously, in full view.

Want to know what we're made of? Again, the cosmic perspective offers a bigger answer than you might expect. The chemical elements of the universe are forged in the fires of high-mass stars that end their lives in titanic explosions, enriching their host galaxies with the chemical arsenal of life as we know it. The result? The four most common, chemically active elements in the universe— hydrogen, oxygen, carbon, and nitrogen—are the four most common elements of life on Earth, with carbon serving as the foundation of biochemistry.

We do not simply live in this universe. The universe lives within us.

That being said, we may not even be of this Earth. Several separate lines of research, when considered together, have forced investigators to reassess who we think we are and where we think we came from. As we've already seen, when a large asteroid strikes a planet, the surrounding areas can recoil from the impact energy, catapulting rocks into space. From there, they can travel to—and land on—other planetary surfaces. Second, microorganisms can be hardy. Extremophiles on Earth can survive wide ranges of temperature, pressure, and radiation encountered during space travel. If the rocky ejecta from an impact hails from a planet with life, then microscopic fauna could have stowed away in the rocks' nooks and crannies. Third, recent evidence suggests that shortly after the formation of our solar system, Mars was wet, and perhaps fertile, even before Earth was.

Collectively, these findings tell us it's conceivable that life began on Mars and later seeded life on Earth, a process known as panspermia. So all Earthlings might—just might—be descendants of Martians.

*

Again and again across the centuries, cosmic discoveries have demoted our self-image. Earth was once assumed to be astronomically unique, until astronomers learned that Earth is just another planet orbiting the Sun. Then we presumed the Sun was unique, until we learned that the countless stars of the night sky are suns themselves. Then we presumed our galaxy, the Milky Way, was the entire known universe, until we established that the countless fuzzy things in the sky are other galaxies, dotting the landscape of our known universe.

Today, how easy it is to presume that one universe is all there is. Yet emerging theories of modern cosmology, as well as the continually reaffirmed improbability that anything is unique, require that we remain open to the latest assault on our plea for distinctiveness: the multiverse.

*

The cosmic perspective flows from fundamental knowledge. But it's more than about what you know. It's also about having the wisdom and insight to apply that knowledge to assessing our place in the universe. And its attributes are clear:

The cosmic perspective comes from the frontiers of science, yet it is not solely the provenance of the scientist. It belongs to everyone.

The cosmic perspective is humble.

The cosmic perspective is spiritual—even redemptive—but not religious.

The cosmic perspective enables us to grasp, in the same thought, the large and the small.

The cosmic perspective opens our minds to extraordinary ideas but does not leave them so open that our brains spill out, making us susceptible to believing anything we're told.

The cosmic perspective opens our eyes to the universe, not as a benevolent cradle designed to nurture life but as a cold, lonely, hazardous place, forcing us to reassess the value of all humans to one another.

The cosmic perspective shows Earth to be a mote. But it's a precious mote and, for the moment, it's the only home we have.

The cosmic perspective finds beauty in the images of planets, moons, stars, and nebulae, but also celebrates the laws of physics that shape them.

The cosmic perspective enables us to see beyond our circumstances, allowing us to

transcend the primal search for food, shelter, and a mate.

The cosmic perspective reminds us that in
space, where there is no air, a flag will not
wave—an indication that perhaps flag-
waving and space exploration do not mix.

The cosmic perspective not only embraces our
genetic kinship with all life on Earth but
also values our chemical kinship with any
yet-to-be discovered life in the universe, as
well as our atomic kinship with the universe itself.

At least once a week, if not once a day, we
might each ponder what cosmic truths lie
undiscovered before us, perhaps awaiting the
arrival of a clever thinker, an ingenious experiment, or an innovative space mission to reveal
them. We might further ponder how those discoveries may one day transform life on Earth.

Absent such curiosity, we are no different
from the provincial farmer who expresses
no need to venture beyond the county line,
because his forty acres meet all his needs. Yet
if all our predecessors had felt that way, the

farmer would instead be a cave dweller, chasing down his dinner with a stick and a rock.

During our brief stay on planet Earth, we owe ourselves and our descendants the opportunity to explore—in part because it's fun to do. But there's a far nobler reason. The day our knowledge of the cosmos ceases to expand, we risk regressing to the childish view that the universe figuratively and literally revolves around us. In that bleak world, arms-bearing, resource-hungry people and nations would be prone to act on their "low contracted prejudices." And that would be the last gasp of human enlightenment—until the rise of a visionary new culture that could once again embrace, rather than fear, the cosmic perspective.

ACKNOWLEDGMENTS

My tireless literary editors over the years these essays were written included Ellen Goldensohn and Avis Lang at *Natural History* magazine—both of whom ensured that, at all times, I said what I meant and meant what I said. My scientific editor was friend and Princeton colleague Robert Lupton, who knew more than I did, in all places where it mattered most. I also thank Betsy Lerner, for suggestions to the manuscript that greatly improved its arc of content.

INDEX

ABOUT THE AUTHOR

Neil deGrasse Tyson is an astrophysicist with the American Museum of Natural History in New York City, where he also serves as Director of the Hayden Planetarium. A graduate of the prestigious Bronx High School of Science, he has a BA in physics from Harvard and a PhD in astrophysics from Columbia. He lives in Manhattan with his family.